Sustainable Communities

Woodrow W. Clark II
Editor

Sustainable Communities

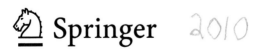
Springer 2010

Editor
Woodrow W. Clark II
Clark Strategic Partners
P.O. Box #17975
Beverly Hills, CA 90209
USA
wwclark13@gmail.com

ISBN 978-1-4419-0218-4   e-ISBN 978-1-4419-0219-1
DOI 10.1007/978-1-4419-0219-1
Springer New York Dordrecht Heidelberg London

Library of Congress Control Number: 2009937583

Thanks to Evelyn Bernstein for the photos that appear between chapters and to Lizabeth Fast for the cover.

Printed on acid-free paper

Springer is part of Springer Science+Business Media (www.springer.com)

Foreword

This book would not have been possible without the dedication and commitment of each of the chapter authors. For some authors, writing a chapter was beyond their "9–5" job, and this book reflects their commitment to sustainability at the local level for their communities. To every chapter author and their staff, friends, and families, thank you. This dynamic and paradigm-changing volume on the topic of sustainable development is focused on communities such as cities, schools, and colleges where the future of our families and children are most at risk. We must act today as each of the chapters represents in their presentations. This book marks a new era: the Third Industrial Revolution.

The new age of the Third Industrial Revolution has been labeled by some as the "green era" or "green economy," but it had already started around the world, especially in Europe and Japan, for over a decade – since the end of the 20th century. More significantly, the book highlights people and communities who have a shared concern and vision along with the will and determination to enact programs and polices that make sustainable development real – not just political rhetoric or "branding" or even the current "buzz word" for obtaining funds and grants. The book presents "The Sequel to an Inconvenient Truth" – actual examples of how communities can and have changed in order to mitigate climate change. Again, thanks to everyone and their colleagues.

Professor Ted Bradshaw, who was the coauthor of *Agile Energy Systems: Global Lessons from the California Energy Crisis* (Elsevier, 2004) needs to share the credit for this book. He died of a heart attack about 2 years after our book was published, so he did not see this volume that was inspired by our work together. Ted's wife, Betty Lou, and their sons have, however. I want them to know and receive the acknowledgement, support, and respect that brought them and me together. May they share in the pride that I have for this book.

I must also thank my two grown children and their spouses, Woodrow W. Clark III (Debra) and Andrea Clark Lackner (David), for their insights, critical comments, and support in my work over the years. In particular, I know that it has not been easy for them, but they both have the intellect and concern that will make them continue to be wonderful children and the best parents.

In particular, I would like to recognize Larry Eisenberg, vice president for Facilities and Planning at the Los Angeles Community College District (LACCD).

While this book was done on my own time, our work together with LACCD over the last 5 years has brought dramatic changes to the nine LACCD colleges and inspired others to do likewise. Frankly, LACCD has become a national and international leader on making colleges (hence communities) energy independent and carbon neutral through renewable energy. This was done with Larry's leadership and support as that of the entire LACCD staff and Board of Trustees. I need to also thank Ms. Jatan Dugar and Russell Vare, who both worked on the final version of the book. Without them, the book would not be completed.

The support from Springer Press and Janet Slobodien in particular has been outstanding. Janet's support from the very beginning of the idea for such a book through the final publication has been remarkable and superb in every way imaginable.

This book is dedicated to my wife, Andrea Kune-Clark, and our son, Paxton Jacob Clark, who will be 2 years old at the time of publication. The book reflects my concern for Paxton. It is his generation to which we baby boomers and soon-to-be-retired elders have left a world that is polluted, quickly turning into an unliveable planet, and causing irreversible health problems.

The book presents actual cases of communities who have recognized these universal problems and done something about it. I hope that communities can use and implement this information for themselves. From what we see and know about the global economic problems, sustainable communities and their "green" development are the future. Hence, the next edition of the book will be four to five times the size of this one, with many more successful cases. It is to this end that I hope to give Paxton and his generation more than just hope, but strong aggressive examples on what they can do. Paxton needs to inherit a world that is extremely different than the one that others and I have created.

Beverly Hills, CA, USA   Woodrow W. Clark II

Contents

Contributors

Lucas Adams JETRO, Los Angeles, CA, USA, lucas_adams@gmail.com

Teresio Asola PIANETA, Settimo Torinese, Italy, teresio.asola@asm-settimo.it

Claire Bonham-Carter EDAM AECOM, San Francisco, California, USA, claire.bonham-carter@aecom.com

Woodrow W. Clark II Clark Strategic Partners, Beverly Hills, CA, USA, wwclark13@gmail.com

Steve Done ARUP Corporation, Los Angeles, CA, USA, steve.done@arup.com

Jatan Dugar University of California, Los Angeles, USA, jatandugar@googlemail.com

Nick Easley Foundation on Economic Trends, Washington DC, USA, neasley@foet.org

Larry Eisenberg Facilities, Planning and Development, Los Angeles Community College District, Los Angeles, CA, USA, eisenblh@laccd.edu

Tony Fairclough Engineering Consultant, tonfair@cox.net

Ken Funaki JETRO, Los Angeles, CA, USA, kentaro_funaki@jetro.go.jp

Andrew Hoffman Los Angeles Community College District, Los Angeles, CA, USA, andrew.hoffmann@build-LACCD.org

Thomas Jensen Enterprise Futures Network, San Franco, CA, USA, tjensen@enterprisefutures.net

Calvin Lee Kwan School of Public Health, University of California, Los Angeles, CA, USA, clkwan@ucla.edu

Natalija Lepkova Vilnius Gediminas Technical University, Vilnius, Lithuania, LT, natalija.lepkova@st.vgtu.lt

Henrik Lund Aalborg University, Denmark, lund@plan.aau.dk

Christine S.E. Magar AIA, LEED AP, MED, cmagar@greenform.net

Brenden McEneaney Office of Sustainability and the Environment, City of Santa Monica, Santa Monica, CA, USA, brenden.mceneaney@smgov.net

Poul Alberg Østergaard Department of Development and Planning, Aalborg University, Denmark, poul@plan.aau.dk

Johanna Partin Renewable Energy Program Manager, Department of the Environment, San Francisco, CA, USA

Bill Radulovich Project EarthRise, Pleasanton, CA, USA, bradulovich@pleasanton.k12.ca.us

Alex Riolfo University of Genova, Genova, Italy, contact through Terssio Asola at ASM, Torino, Italy, teresio.asola@asm-settimo.it

Christine Meisner Rosen Haas Business School, University of California, Berkeley, Berkeley, CA, USA, crosen@haas.berkeley.edu

David Schoenberg Enterprise Futures Network, Haas School of Business, University of California, Berkeley, CA, USA

Patrick Stoner Program Director, Local Government Commission, Sacramento, CA, USA, pstoner@lgc.org

Phil Ting Assessor-Recorder of San Francisco, San Francisco, CA, USA

Russell Vare Policy Advisor Consultant, russellvare@gmail.com

Wang Weiyi Jiaxing University, Zhejiang, P. R. China

Li Xing Aalborg University, Aalborg, Denmark, xing@ihis.aau.dk

Stephanie Whittaker Sustainable Development Advisor, AECOM, San Francisco, CA, USA, stephanie.whittaker@aecom.com

Chapter 1
Introduction and Overview

Woodrow W. Clark II and Russell Vare

The "Green Revolution" has begun globally.

Starting in the international academic world as "sustainable development" and identified in California as one of the four key economic development areas [1], the "Green Revolution" began over two decades ago with the UN Brundtland Report. Sustainable development became known as the "Green Revolution" [2] and then to the general public through Al Gore's "An Inconvenient Truth" winning both an

W.W. Clark II (✉)
Clark Strategic Partners, Beverly Hills, CA, USA
e-mail: wwclark13@gmail.com

W.W. Clark II (ed.), *Sustainable Communities*, DOI 10.1007/978-1-4419-0219-1_1,   1
© Springer Science+Business Media, LLC 2010

Academy Award and a Noble Peace Prize in 2007. However, now the concept and meaning of "green" have been confused with "clean," "alternative," and "nuclear power" among other terms describing the solutions to energy and the environmental climate change [3].

Climate change and global warming have been acknowledged on many levels – increasing public awareness, business green marketing campaigns, new government legislation such as California's "AB32 The Global Warming Solutions Act," Vice President Al Gore's film "An Inconvenient Truth," and the Nobel Peace Prize awarded to the Intergovernmental Panel on Climate Change in 2007. Researchers and political decision makers (including the new president of the USA, Mr. Obama) around the world have now recognized the need to take actions.

Communities are entering a world that some authors have now labeled "The Third Industrial Revolution" [4], as society moves from fossil fuels (the Second Industrial Revolution) to renewable energy generation, technologies in sustainable, smart communities. More common terms include green technology, renewable energy, earth friendly, leadership in energy, environment and design or LEED building standards, eco-conscious, recycling, and organic products and services. Each of these terms and concepts touches on aspects of sustainability. It is important to remember that sustainability includes more than just carbon emissions, solar electricity, or recycling. It includes all aspects of living in order to preserve our communities for future generations.

Renewable energy is a key component of sustainability not only for buildings, but also for communities such as infrastructures. A deeper examination of sustainable living illuminates how interconnected our world is, and the energy required to keep it moving. For sustainability, so much potentially must be changed – possibly everything. It is daunting to think about how to change everything from our buildings to our transportation systems, just in order to keep our planet hospitable.

Sustainability is achievable. It can be done, and must be done, at the community level. Block by block, city by city, region by region, communities can change how they live. The electric car in the above figure is the first hybrid (the Lohner-Porsche carriage) from Germany in 1903. There is nothing new about either electric or hybrid cars. They simply have not been commercially made due to economic and political pressures from the oil and car companies in favor of fossil fuels and internal combustion engine manufactures.

This book gives concrete examples of how communities all over the world are making changes in themselves toward sustainable futures. The use of old technologies that existed in the Second Industrial Revolution (fossil fuels and nuclear power) but were not commercialized is now becoming popular due to the public awareness of global warming and the dramatic changes in the climate and hence weather throughout the world. The use of hybrid technologies is providing a short transition period (from 5 to 10 years) until the next generation of all electric, hydrogen, and other environmentally sound vehicles are in the marketplace. That next generation is occurring now in 2009.

Fig. 1.1 Porsche All Electric Car, 1903 (an example of how the Third Industrial Revolution was delayed over a century ago)

The Toyota Prius and Honda Civic, Camry, and now Clarity are all examples of the move into the Third Industrial Revolution. The Prius, for example, with its combination of electric and gasoline motors did just that by demonstrating constant 45 mph gas mileage, reliability, and performance. The USA and many EU automakers did not fully realize the consumer demand for environmentally sound personal transportation, let alone mass transit and shipping needs for vehicles, trains, and ships.

Meanwhile, the public sector has taken the lead in making communities sustainable. And as in the next chapter on providing an environmental and historical background context for urban pollution over the last century notes, the Second Industrial Revolution did irreparable harm to the environment, health, and development of most urban communities around the world. The private sector was responsible along with public sector compliance and even encouragement. However, today companies such as Google are installing solar systems on all their headquarters buildings and researching electric vehicles and developing software to accompany smart-grid monitoring for home computers (Chapter 8). Others are achieving LEED levels for certification.

The Los Angeles Community College District is one of the most aggressive public sector systems as it recycles trash and waste, but also conserves resources and is more efficient with its use of water, waste, and energy. Even more significantly, LACCD is achieving gold and higher levels of LEED while making all its nine campuses energy independent and carbon neutral (Chapter 2). Frederikshavn, Denmark (Chapter 10), is incorporating multiple renewable energy technologies strategically over time to make the entire city energy independent. Communities in California and UK are redesigning space to increase pedestrian and bicycle traffic to reduce the amount of fuel burned in vehicle use (Chapters 3, 4, and 9). Furthermore, Chapter 7 focuses on the achievements of one U.S. public school in Northern California to become sustainable.

Communities in China and Spain are totally sustainable through public policies that support renewable energy power generation and hence sustainable communities, like the German Feed-in-Tariffs (established in the early 1990s, but expanded in early 2000), which fixed rates in order to provide rebates to consumers and communities in general. Spain has now done the same. China has 5-year national plans that help it "leap frog" technologies that mitigate environmental problems. Other communities in Japan have been sustainable for many years, since Japan as a nation must either import all its energy or generate it on the island nation itself. American city governments, such as San Francisco and Santa Monica in California, are leading by example, supporting LEED-designed buildings, renewable energy generation, and storage devices (Chapters 5 and 6). Other nations such as Italy (Chapter 11) have been active in sustainable regions, while Lithuania (Chapter 12) has begun a national focus on sustainability.

The purpose of this book is to provide advanced undergraduate and graduate students, in a variety of disciplines, specific background material and cases of sustainable communities, along with the technical tools to make sustainable development a reality. This book includes cases from the USA, Europe, and Asia that give specific actions taken to become sustainable. In most cases, these communities are now off the power grid or will be so in the near future. As expected, many of these cases look at local governments and cities that have taken successful concrete steps to implement sustainability programs. This book also includes other types of communities such as school districts, business districts, and shopping malls that are implementing sustainability programs at the micro-level.

The book provides a vast, but not the total, amount of materials and data. The intent is to give students, teachers, scholars, policy makers, and practitioners a perspective and an understanding of sustainable development in communities, citing actual cases along with some well-established resources and tools. Some experts call these "smart grid communities (city)," but it is far better to refer to them as both sustainable and smart so that both ideas of infrastructures (from water to telecom) are included in renewable energy power generation along with LEED buildings standards.

The International Perspective

California is often considered an unusual case. However, think about August 2003 in the USA. Over 50 million Americans in the upper mid-west and eastern parts of the USA lost their power for a few days. Not just for a few minutes but for many days. Over 3 million Canadians also suffered from this massive energy blackout. The impact of that blackout mirrored a similar series of energy programs throughout Europe during the summer of 2003. France, for example, attributed at least 12,000 deaths to the shortage of energy. And yet the country has over 75% of its energy produced by nuclear power plants.

On the other hand, Japan had entered into the new green economic paradigm a decade or two before that, in the 1970s, due to its historic need to get off being dependent on foreign oil and gas. The history behind Japan and World War II was primarily based on its need and demand for increased fuels for its industry. In 1973, the global energy crisis once more brought that problem to the Japanese leadership.

Japan began seeking environmentally friendly technologies for its own energy power uses for buildings and vehicles. Hence, the hybrid car was born in the early 1990s in Japan, recognizing that greater fuel efficiency along with higher mileage per gallon would mean less use of fossil fuels. With government support and encouragement, the Toyota Prius was built from licensed USA-developed "regenerative" braking technology, rejected by the American Big Three carmakers in the mid-1990s. Toyota licensed the technology and put it into demonstration and then commercial use by the late 1990s.

The first demonstration models were introduced into the USA in the early part of the 21st century, especially in Southern California where driving and the costs of fuel were particularly acute along with the concern for the environment. Japan has even taken the next step in developing small-scale hydrogen fuel cells to power small buildings. Currently, there are just over 2,000 homes powered by these small stations, with the goal that by 2020, a quarter of all the homes in Japan will be powered by fuel cells (Aki). Currently, hydrogen for these fuel cells is produced from natural gas, but renewable energy electrolyzers are being developed and demonstrated for hydrogen fuel cells.

Back in Europe about the same time (early 21st century), Germany recognized its need to get off fossil fuels. Historically again, Germany fought wars over its need for energy to power its factories. Policy makers and leaders saw the need to dramatically increase renewable energy power generation to get the nation off its growing dependency on fossil fuels. By the turn of this century, almost all of Germany's energy power came from natural gas (primarily from Russian pipelines) and another third from coal-fired plants. That and the use of hydroelectric power meant Germany needed to change rapidly into the new paradigm. And it did. The 9/11 attacks on the USA further confirmed the need for the government to take actions.

Through the creation of the German "Feed-in Tariffs," which were laws or set payment for higher prices of power generation by consumers, the nation started to subsidize also the economic and employment growth of renewable energy power supplies. This was not neoclassical supply-side economics. This was where the government recognized the problem of fossil dependency and its impact on the environment and took action. Spain was soon to follow. The net result less than a decade later is that Germany is the largest global supplier of solar systems and Spain is second. Furthermore, both countries are world leaders in wind power generation. And they are not alone; Denmark now has about 40% of its power generation from wind, which will be over 50% by 2015. Other European nations are following suit.

The People's Republic of China (PRC or China) presents a remarkable modern-day dilemma for the USA. The nation held the 2008 Olympics under incredible conditions and scrutiny. While much of the criticism from air pollution to individual human rights was justified, the PRC set the stage for the rest of the 21st century. The USA, for example, just lost to the PRC the distinction of being the number 1 world air emission polluter. China is now #1. That is not something to be proud of, but if calculated on a per capital basis, the USA would still rank #1. The reason why that distinction is important to the USA and the rest of the world is that the PRC will certainly reverse the statistics and within a matter of 3–5 years will be well below the USA and other Western nations. How?

The why is important: China does not want to be the world's worse polluter. But more importantly, the PRC has the ability to "leap frog" the mistakes of the West in the development of its energy infrastructures [5]. Under the Five-Year Central Government Plans, the PRC can state plans, provide financial resources, and set goals to which the nation can develop its sustainable resources. The current 11th Five-Year Plan does not provide any discussion of sustainable development but most likely will have it discussed at length in the 12th Five-Year Plan. The importance is that China will be able to articulate and then execute a sustainable energy plan.

The plan would be well financed and focused on the nation to reduce its carbon footprint, which causes climate change and global warming. What will be interesting is if the Chinese leap frog over the Western nations who developed central power plants into agile sustainable energy infrastructures and local on-site power systems using renewable energy for homes, office buildings, resorts, shopping areas, and public buildings like colleges, universities, and government facilities.

Furthermore, contrary to the West, China has developed a government-oriented economy. However, in recent years, there has been more and more of a "market push," which may be called into serious question, given the global economic crisis of 2008–2009.

The new energy paradigm throughout the world will continue to evolve slowly in the rest of the world because of past economic and political history. The Cold War ended in the 1990s, only to be revisited in the Middle East and Eastern Europe, but also the entire way in which the world viewed economic growth and its impact on the climate turned around dramatically.

The old supply-side economic paradigm embraced by the UK Prime Minister Thatcher and the US President Reagan imploded [3], as the world and local regional economics felt the shock waves from the privatization and deregulation of basic infrastructures from water and waste to education and health as well as transportation and energy. Nation after nation mistakenly had begun to turn over these basic infrastructure sectors to the private business sector. China now appears to recognize this economic problem and does not have to make the same mistakes of the West or rise out of their aftermath.

There is no need to review and analyze once more what happened in each of the Western nations about their infrastructure sectors or how in the end, the government and the public had to save the industry(s). Clark and Isherwood (2007) did that for the Asian Development Bank with a focus on Inner Mongolia (IMAR) in China. The report recorded the history of Western energy infrastructures and their impact on the environment and how the IMAR could "leap frog" the mistakes of the West.

Today the evidence is obvious as to the environmental consequences in energy, transportation, education, and health services, among others, including the 2008 subprime home-loan debacle. Competition for profit emerged whereby fewer companies had replaced public monopolies with private ones in control of vital infrastructure sectors. The general public suffered the most. Energy infrastructures throughout the world are classic examples of what happened with the old paradigm of deregulation and privatization.

Agile sustainable communities and smart networks are a significant part of the new energy paradigm [6]. Public organizations and private companies that include energy producers, generators, transmission lines, and distributors must now focus on specific local consumers such as colleges, shopping malls, walking streets, residential and retirement communities, and office complexes. Agile energy plays one of the key roles in all communities of whatever size. Often ignored or taken for granted, energy is a critical infrastructure in any community, region, or nation.

Consider what New Zealand could have done. The country upgraded transmission lines at a cost of $600 million dollars in order to increase electrical transmission by 200 MW. Alternatively, at $3.00/W, localized solar PV could have been installed, resulting in the same cost as upgrading the transmission lines. However, after taking into account the savings associated with purchasing less fuel and that technologies inevitably improve resulting in lower solar PV prices, installing solar PV becomes more enticing and economically justified when compared to building more transmission lines.

Agile Sustainable Systems

The energy needs for all communities around the world are growing more complex each year. Population increases, cities expand, and power demands climb. Pollution in the air and water has become serious, causing health problems for young and old residents. Local and national governments around the world are now implementing carbon dioxide regulations in order to stop pollution or at least try to reverse it. The challenge of supplying energy for increasing demand, while reducing carbon emissions, calls for more complex and creative solutions. It requires taking into account efficiency, renewable energy, and entirely new systems to change the way people live and the way they think about using electricity and energy in general. The "Sequel to an Inconvenient Truth" means stopping and reversing pollution of all kinds.

Energy is one of the key areas for making communities sustainable and pollution free.

Renewable energy is a large part of the solution. The key to green energy production are local distributed generation (DG) systems and on-site power generation that are "agile," or flexible to meet the local needs. While a central power grid exists for many consumers, local renewable energy generation allows consumers to generate their own power.

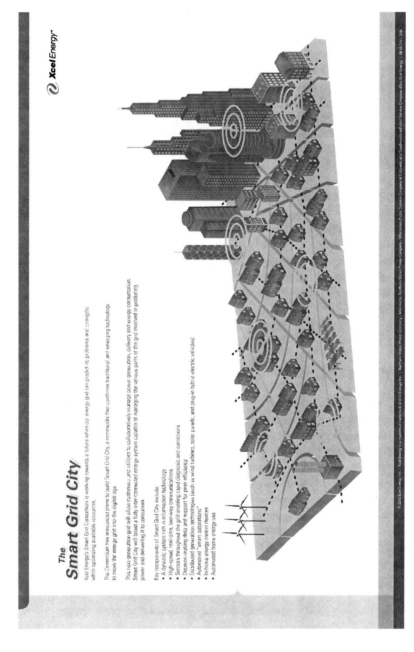

Fig. 1.2 The Smart Grid (with homes, offices, and shopping areas within the same region)

Fig. 1.3 The Traditional Central Power System (as developed in the early 20th Century but the basic model now into the 21st Century)

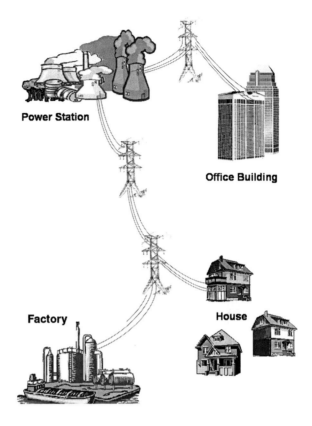

Power Station

Office Building

Factory

House

Central grids were developed as part of the Second Industrial Revolution in order to supply energy demand for consumers. Usually these systems required long distances for transmission lines, pipelines, shipping, and other means of transmission for the energy. The standard approach was usually to do the capital costs for the actual central plant with its processing of raw materials but not the transmission costs that were added on. The central plants were also built to process fossil fuels such as coal, oil, and natural gas. Nuclear power plants created the same problems and cost issues.

Note: The charts above illustrate that on-site power is the ability for a building (home, office, or business) and clusters of buildings to have power from renewable energy such as solar, wind, geothermal, ocean, or wave power. These technologies and systems are the core of the Third Industrial Revolution. Other new technologies are being developed as well as storage devices such as lithium-ion batteries, fuel cells, and others to store the energy at night when the sun is not shining or the wind not blowing. Agile energy systems and hence communities combine or provide the flexibility to combine central power grid supply with local on-site renewable energy generation from natural resources.

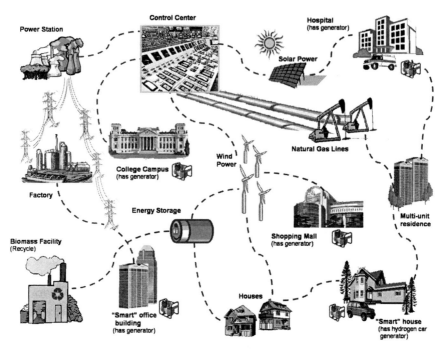

Fig. 1.4 The Agile Power System Model (which continues to have a central grid but now local communities also generate and store energy from renewable power sources)

California is a good example of where small communities are now producing their own energy from renewable sources. As such it is a model for each region or community to generate its own power from renewable sources such as solar, wind, geothermal, ocean, or some combination of all of them. Another local energy generation strategy are hybrid technologies such as solar and PV systems, while in the mountains there may be runoff of the river water flow or near the ocean tides. California already has considerable amounts of geothermal power as well as significant wind farms. However, more local on-site wind generation is possible. In all these cases, for example, hydrogen can be produced from renewable or green sources and then stored close to the needs and demands of communities.

Consider how energy systems are evolving today. In the book *Agile Energy Systems: Global Lessons from the California Energy Crisis* (Elsevier, 2004), Clark and Bradshaw document this point (see http://www.elsevier.com). Agile energy systems are a combination of local or regional energy systems (including on-site and self-generation) and central grid systems. In the future, the central grid can be used primarily for redundancy and back-up purposes and in some cases like a battery for storage of energy when the sun is not shining or the wind not blowing. Hence, these agile systems are not particular technologies or market mechanisms, but rather a new paradigm that is part of the Third Industrial Revolution.

Sustainable Community Focus

The book shows that solving these problems can come from the ground-up or local level that makes big changes. Often communities start and implement change. Communities around the globe are tackling these challenges, with case studies from the USA, Europe, and Asia highlighted in this book. Each chapter examines communities that are looking at these issues of increased energy and reduced emissions.

A challenge as large as global warming and climate change will require involvement from everyone. Building on the success of other communities around the world has implemented sustainability practices that prove that sustainability can be achieved. And that it must be achieved by the participation of the community. People are at stake here in terms of their personal health, but also of their community and regions. Change starts at home.

References

1. Clark II, Woodrow W. and Todd Feinberg, "California's Next Economy", Governor's Office of Planning and Research, Sacramento, CA. Local Government Commission web site at http://www.lgc.org January 2003, pp. 1–56.
2. Clark II, Woodrow W. and Jeremy Rifkin, et al., "Hydrogen Energy Stations: Along the Roadside to an Hydrogen Economy", Utilities Policy, Elsevier Press, January 2005 (13), pp. 41–50.
3. Clark, Woodrow W. II and Michael Fast. *Qualitative Economics: Toward a Science of Economics*. London, UK (Coxmoor Press 2008).
4. Rifkin, Jeremy, 2004. *The European Dream: How Europe's Future is Quietly Eclipsing the American Dream*. Tarcher, Pengin. See also the Foundation for Economic Trends at: www.foet.org.
5. Clark, Woodrow W. II and William Isherwood, "Energy Infrastructure for Inner Mongolia Autonomous Region: five nation comparative case studies", Asian Development Bank, Manila, PI and PRC National Government, Beijing, PRC, 2007. To be published in Utility Policy Journal, Spring 2009.
6. Clark, Woodrow W. II and Larry Eisenberg, "Creating Agile, Sustainable and Smart College Campuses: The Case of the Los Angeles Community College District", Utility Policy Journal, June 2008.

Chapter 2
The Role of Business Leaders in Community Sustainability Coalitions: A Historical Perspective

Christine Meisner Rosen

The environmentalist Paul Hawken recently published an article in *Orion Magazine* in which he lauded the hundreds of thousands of organizations in the USA and countries around the globe that are working to find solutions to an array of increasingly serious environmental, economic, and social justice problems. These groups come from all parts of civil society, including business. As Hawken puts it, they include "research institutes, community development agencies, village- and citizen-based

C.M. Rosen (✉)
Haas Business School, University of California, Berkeley, CA, USA
e-mail: crosen@haas.berkeley.edu

W.W. Clark II (ed.), *Sustainable Communities*, DOI 10.1007/978-1-4419-0219-1_2,   13
© Springer Science+Business Media, LLC 2010

organizations, corporations, networks, faith-based groups, trusts, and foundations." They are part of broad-based, grassroots movements that are working, often collaboratively, at the local as well as national and international levels, to restructure communities and economies to ameliorate poverty and avert the looming social and economic crises that will result from – or be worsened by – the great environmental problems of our times: climate change, toxic air and water pollution, and the depletion of natural resources.[1]

This chapter will explore the role that progressive business leaders played in late nineteenth and early twentieth century movements to regulate urban coal smoke – that era's most threatening form of air pollution – in order to draw attention to the similar, potentially equally important role that business leaders may be able to play in today's community sustainability movements. The anti-smoke movement focused on developing technological solutions to the smoke problem and on enacting regulations to force the owners of smoking furnaces and boilers to use these technologies to abate their smoke. The regulations specifically targeted business. Like today's community sustainability movements, the anti-smoke movements involved a variety of interest groups. Like today's movements, they were locally organized and focused, though their leaders too communicated with each other regarding regulatory strategies and abatement technologies, through letters, consults, and meetings of professional organizations. Most significantly, like today's groups, they often engaged in long and strenuous political and legal battles over regulatory policy.

What is remarkable about the early struggles over smoke pollution regulation is that business leaders were among the most important and influential leaders in the struggle – despite the fact that the regulations targeted the business community. This chapter will argue that while their leadership role is now largely forgotten, it has great relevance for today's environmental struggles. Working alone, or more often in collaboration with women, public health, and other Progressive Era urban civic reformers, groups of reform-minded urban business leaders helped secure most of the anti-smoke crusade's regulatory successes. Their accomplishments speak to the strategic desirability – even necessity – of business participation in and shared leadership of today's movements for community sustainability.

Business Leadership of Urban Anti-smoke Movements

Coal smoke became an increasingly serious problem in most American cities after the Civil War. The burning of increasingly massive quantities of soft, cheap bituminous coal for transportation as well as industrial, commercial, and residential power and heat put a dark pall of heavy black smoke over these places. In contrast to the invisible green house gases (GHGs) responsible for climate change, coal smoke was a highly visible, palpable problem in late-nineteenth-century cities. The smoke not

---

[1] Paul Hawken, "To Remake the World," *Orion Magazine*, www.orionmagazine.org/index. php/articles/article/265May/June2007.

only blew into the homes of residents, leaving scums of black filth on furniture and laundry drying on outdoor (and indoor) clotheslines, but also swept into downtown offices, choking the bankers, lawyers, accountants, clerks, and others at work, covering their clothing and office supplies with soot. It also swept into downtown stores, dirtying the products on public display, and made its way into warehouses, where it damaged goods in inventories. It also led to respiratory and other illnesses.[2]

Public disgust and anger at the choking, filthy smoke led to an explosion of anti-smoke activism on the part of health reform and women's organizations as well as a widening array of business organizations – from Chambers of Commerce to Citizens' Associations and engineering organizations to neighborhood improvement associations and business-led political reform groups. In some instances, these business-led groups were the primary, or even the only, leaders of the anti-smoke campaigns. In most cases, however, they were the allies of the women and public health reformers, helping them in ways that were absolutely essential to the success of smoke reform.

My first example of business leadership comes from Chicago. It concerns the Chicago Citizens' Association, an organization of businessmen who were dedicated to reforming political and environmental conditions in the city. During the late 1870s and 1880s, its members studied, proposed regulatory solutions for, and lobbied for legislation to deal with several air, noise, and water pollution problems, in addition to a wide range of the city's many other physical, social, and political ills.[3]

After helping reform-minded public officials, newspapers, and residents win a big victory in the city's so-called stench wars against the foul emissions of its packing and rendering house industries in 1878,[4] the leaders of the Chicago Citizens' Association mobilized to secure the passage of an ordinance to abate the city's growing problems with black coal smoke. They succeeded in persuading the city council to pass a smoke ordinance in 1881 that required commercial building owners to install smoke-abating equipment on their coal-burning furnaces and steam boilers to reduce their emissions of heavy black smoke to specified levels. Though it was

---

[2]For descriptions of urban smoke problems, see David Stradling, *Smokestacks and Progressives: Environmentalists, Engineers, and Air Quality in America, 1881–1951* (Baltimore and London: The Johns Hopkins University Press, 1999). This kind of air pollution was a problem wherever coal was used as fuel on an industrial scale: see Frank Uekoetter, *The Age of Smoke: Environmental Policy in Germany and the United States, 1880–1970* (Pittsburgh: University of Pittsburgh Press, 2009); Peter Thorsheim, *Inventing Pollution: Coal, Smoke, and Culture in Britain Since 1800* (Athens: Ohio University Press, 2006).

[3]For an overview of the Citizens' Association's attempts to reform machine politics in Chicago, see Richard Schneirov, *Labor and Urban Politics: Class Conflict and the Origins of Modern Liberalism in Chicago, 1864–97* (Urbana and Chicago: University of Illinois Press, 1998), 53–63. Citizens' Association of Chicago, *Annual Reports of the Citizens Association of Chicago* (Chicago, 1876–1925) contains the record of its many initiatives in the area of environmental and civic improvement.

[4]Christine Meisner Rosen, "The Role of Pollution Regulation and Litigation in the Modernization of the U. S. Meat Packing Industry, 1865–1880," *Enterprise and Society: The International Journal of Business History*, June 2007: 327–8.

not the first municipal smoke ordinance in the U.S. history, this law appears to have been the first that city officials actually enforced. It authorized the Department of Health to impose fines of $10–$50 on violators. Following the precedent it had set in the battle against the packing and rendering house stench nuisance, the Citizens' Association members helped the department identify and prosecute violators.[5]

Although the association's efforts led to several years of apparently quite remarkable improvements in air quality, its victory was short lived. Enforcement of the ordinance declined, due to a lack of inspectors to monitor compliance and the refusal of local juries to convict violators. As a result, the Citizens' Association lost interest in the issue.

In 1892, however, a new business-led reform organization, the Smoke Prevention Society (SPS), stepped into the gap, going even further than the Citizens' Association had to help the city enforce its smoke laws in preparation for the 1893 World Columbian Exposition. Among other things, SPS leaders hired a staff of engineers (at their members' expense) to inspect smoking boiler plants and advise their owners about how best to control the smoke. When a hard core of property and tugboat owners continued to balk at coming into compliance with the 1881 smoke regulations, they also hired their own attorney to prosecute violators. They successfully prosecuted violators until a group of tugboat owners and railroads and other particularly stubborn repeat violators began insisting on jury trials.[6]

The reluctance of juries to return verdicts against defendants led to a series of legal defeats, and the SPS too withered away. Within a few years, however, reform-minded businessmen and technical experts were at it again, calling on commercial property owners to abate their smoke, publicizing the availability of technologies to do this, and pressing for better enforcement of the smoke ordinance. In 1903 they helped secure the establishment of a specialized smoke inspection agency, helping staff it with increasingly professionalized engineers who helped building owners diagnose their smoke problems and find technical solutions to them.[7]

These Chicago smoke reform movements exemplify how business reform organizations instigated anti-smoke crusades, mobilizing public opinion against the smoke

---

[5]Chicago Department of Health, *Annual Report for 1881 and 1882* (Chicago, 1882), 31–2, 34–44; Citizens' Association of Chicago, *Annual Report for 1878* (Chicago, 1878), 8; Citizens' Association of Chicago, *Annual Report for 1882* (Chicago, 1882), 6–7; Citizens' Association of Chicago, *Annual Report for 1883* (Chicago, 1883), 6–8; Citizens' Association of Chicago, *Annual Report for 1885* (Chicago, 1885), 5–6; Citizens' Association of Chicago, *Annual Report for 1885*, 15; Citizens' Association of Chicago, *Annual Report for 1887* (Chicago, 1887), 13; Citizens' Association of Chicago, *Annual Report for 1888* (Chicago, 1888), 25–6; Citizens' Association of Chicago, *Annual Report for 1889* (Chicago, 1889), 8–9. Bitterly opposed by many businesses, the smoke ordinance was upheld by the Illinois Supreme Court in 1884 in Harmon v. City of Chicago, 110 Ill. 400, 51 Am. Rep. 698.

[6]Christine Meisner Rosen, "Businessmen Against Pollution in Late Nineteenth Century Chicago," *Business History Review*, Fall 1995: 351–97.

[7]Ibid., 398; Harold L. Platt, *Shock Cities: The Environmental Transformation and Reform of Manchester and Chicago* (Chicago: University of Chicago Press, 2005), 479–80; Frank Uekoetter, *The Age of Smoke*, 31–3.

problem and helping municipal officials pass and enforce smoke regulation with little or no support from other civic organizations and interest groups. They also reveal the up again, down again, two steps forward, one step (or more) back nature of the struggle, so familiar to modern environmentalists. A group would do battle against the smoke problem, accomplishing some success, only to be thwarted by an adverse legal decision or other problems. After the defeat, local anti-smoke reformers would regroup and begin a new movement to eliminate the smoke problem, continuing the battle.

The exclusively business-led anti-smoke movements that took place in Chicago during the 1880s and 1890s were the exception, however, rather than the rule in the late-nineteenth- and early-twentieth-century smoke wars. More often anti-smoke business leaders and their organizations worked in collaboration with women's groups and other non-business interests to achieve their regulatory goals. In some places, women organized to support regulatory movements begun by their city's leading businessmen. In other instances, the women took the lead and the men became their allies. In either case, this activism usually proceeded in waves, with a variety of business and non-business organizations working in collaboration with each other over time. Typically one group would begin agitating for smoke reform. One or more additional organizations would join in the effort to pass regulation. This activity would continue until their city council enacted a smoke ordinance. All would be well, until enforcement problems developed, often due to adverse legal decisions, or frustration set in as other problems with the existing ordinance became clear. Then, as in Chicago, a new surge of anti-smoke activism would begin. The cycle often repeated several times, sometimes continuing for as many as 30 or 40 years.[8]

In St. Louis, for example, a group of business and government leaders commenced a broad community movement for smoke regulation in 1891 by convening a large meeting of "prominent citizens, representing fifteen city clubs and commercial bodies" to discuss the smoke problem and find a solution to it. This group appointed a committee of engineers who developed the smoke ordinances passed in 1893, the first enacted in St. Louis, which required businesses to abate their emission of "dense black and thick gray," while creating the city's first smoke regulation agency (the St. Louis Smoke Commission). Though spearheaded by men, women reformers quickly involved themselves in the campaign. The members of the Wednesday Club, an elite women's club, helped the men mobilize public support for the proposed ordinance and lobby city council members. After the ordinances were enacted, Wednesday Club members also helped officials in the newly created

---

[8]For general overviews of these movements in the USA as a whole, see Stradling, *Smokestacks and Progressives*; Uekoetter, *The Age of Smoke*; Robert Dale Grinder, "The Battle for Clean Air: The Smoke Problem in America, 1880–1917" (Ph.D. dissertation, Department of History, University of Missouri, 1973); Robert D. Grinder, "The Battle for Clean Air: The Smoke Problem in America, 1880–1917," in *Pollution and Reform in American Cities, 1870–1930*, ed. Martin V. Melosi (Austin & London: University of Texas Press, 1980), 83–103.

Smoke Abatement Department enforce their provisions by deploying members to serve as observers who reported on violators.[9]

According to historians Joel Tarr and Carl Zimring, the "ordinances appear to have been relatively successful in reducing some of the worst smoke nuisances." However, in 1897, the Missouri Supreme Court invalidated them. After this legal debacle, community groups began a new campaign for smoke regulation. They secured state-enabling legislation in 1901 that enabled the city to enact a new, but weak, smoke ordinance. In response to continued community agitation and a legal decision validating the new ordinance, the council passed a stronger law in 1904, but refused to appoint a smoke inspector willing to vigorously prosecute violators.[10]

The community campaigns continued, led by the Civic League of St. Louis, a business reform group, and the women of the Wednesday Club. While the Civic League focused on proposing and lobbying for structural reforms of the office of the smoke inspector, the women again tackled the problem of enforcement, sending members of the newly formed group called the "Women's Organization for Smoke Abatement" out as observers to identify and report violators of the poorly enforced smoke ordinance to the city's Smoke Abatement Department and the newspapers and lobbying for prosecution of the perpetrators. As their movement developed momentum, other civic groups began to join in, including the Businessmen's League, the Million Population Club, the Socialists, and most of the city's newspapers. In 1911, a wealthy local philanthropist, Colonel James Gay Butler, spent $50,000 to hire six inspectors and a lawyer to help the city prosecute offenders, a development that apparently shook things up enough to force the city to finally institute many of the structural reforms desired by the activists. The women activists continued to deploy their members to watch towers to help the city's smoke inspectors identify violators. These efforts reduced, though they did not eliminate the city's smoke nuisance.[11]

When World War I began, the nearly 25-year-long anti-smoke community crusade against smoke ground to a halt and the smoke problem worsened. The crusade resumed in 1923, when anti-smoke leaders of the Women's Organization for Smoke Abatement joined forces with anti-smoke members of the Chamber of Commerce to form the mixed gender Citizens' Smoke Abatement League. The league conducted experiments with smoke abatement technologies and smokeless fuels, worked with

---

[9]Lucius H. Cannon, *Smoke Abatement: A Study of the Police Power as Embodied in Laws, Ordinances and Court Decisions* (St. Louis: St. Louis Public Library, 1924), 211–2, 222–3; Joel A. Tarr and Carl Zimring, "The Struggle for Smoke Control in St. Louis," in *Common Fields: An Environmental History of St. Louis*, ed. Andrew Hurley (St. Louis: Missouri Historical Society Press, 1997), 203–4; Robert Dale Grinder, "The War Against St. Louis's Smoke 1891–1924," *Missouri Historical Review* 69 (January. 1975): 192–3.

[10]Tarr and Zimring, "The Struggle for Smoke Control in St. Louis," 204; Cannon, *Smoke Abatement*, 213–22; Grinder, "War Against St. Louis's Smoke," 193–4.

[11]Grinder, "War Against St. Louis's Smoke," 194–205; Cannon, *Smoke Abatement*, 222–5; Mrs. Ernest R. Kroeger, "Smoke Abatement in St. Louis," *American City*, June 1912, 907–8.

the city's Division of Smoke Regulation to help ordinary citizens as well as business reduce their emissions, raised hundreds of thousands of dollars to educate the public about how to abate smoke, and lobbied for more power for smoke inspectors, tough controls on the types of coals that could be sold in the city, and other changes to improve smoke regulation. When its members' energy began to lag in the late 1930s, the St. Louis *Post-Dispatch* began a new anti-smoke campaign that led to the formation of a new anti-smoke committee composed entirely of local business leaders, in 1939. It succeeded in getting enacted a new smoke ordinance that required all coal users, manufacturers, commercial businesses, railroads, and residents alike to use mechanical stokers and smokeless fuels.[12]

In other cities, such as Cincinnati and Pittsburgh, where women reformers instigated smoke reform, the more politically astute leaders soon realized they needed the support and political clout of business leaders and reached out to them. Sometimes they were able to do this relatively easily. In other cases, however, they worked for years before they succeeded in forging effective coalitions with business leaders.

In Cincinnati, for example, the upper class members of the Women's Club, with the help of some local public health reformers, began a movement in 1904 to force the city to enforce its almost 25-year-old, but long dead, letter smoke ordinance. They quickly realized that they lacked the power to wield enough influence on their own, and so, with the help of their personal and class connections, they began enlisting the support of prominent businessmen. By 1906 they had enough such support to form the mixed gender Smoke Abatement League. The league included powerful business leaders as members and financial backers, including the presidents of Proctor & Gamble, Stearns & Foster, American Smelting & Refining Co., and Cincinnati Milling Company as well as Charles P. Taft, the editor of the *Cincinnati Times-Star*. Like anti-smoke reformers in Chicago and St. Louis, league members investigated the city's smoke problems and made citizens' arrests of offenders, while deploying their more connected members' considerable political clout to lobby for the strong enforcement of the 1881 smoke ordinance. The league also hired an engineer to help it with its investigations and help the owners of smoke boiler plants come into compliance with the 1881 smoke ordinance. The city eventually appointed this man as its chief smoke inspector. To help him enforce the law, the league deployed members to watch towers to identify and document violations. The league worked closely in this way with the Smoke Department through the 1910s and 1920s, before fading away during the Great Depression.[13]

---

[12]Tarr and Zimring, "The Struggle for Smoke Control in St. Louis," 205–20; Stradling, *Smokestacks and Progressives*, 163–76; Uekoetter, *The Age of Smoke*, 72, 77–82; Cannon, *Smoke Abatement*, 225.

[13]Stradling, *Smokestacks and Progressives*, 52–5, 211, n. 40; Uekoetter, *The Age of Smoke*, 72. See the lists of "Subscribers and Donors" in the league's annual reports for names of individuals and corporate members. For a first person description of how league members helped the city smoke inspector enforce the law, see Matthew Nelson, "Smoke Abatement in Cincinnatti," *American City* 2 (January 1910): 8–10, Reprinted in H. Wayne Morgan (ed.), Industrial America: The Environment and Social Problems, 1865–1920 (Chicago: Rand McNally College Publishing Company, 1974), 73–77.

Women were also the movers and shakers behind smoke reform in Pittsburgh for many years. The Ladies' Health Protective Association (LHPA), an upper class women's organization founded in 1889, began to campaign for the institution of an enforceable smoke ordinance in 1891. They, too, quickly realized they needed the help of local business leaders to make significant headway in their battle. Despite their personal connections (through marriage) with the city's industrial elite, however, they faced a much more uphill battle trying to bring businessmen into their nascent movement than the Cincinnati women, due to the strong opposition of Pittsburgh's iron and other manufacturers. Their 1892 effort to solicit the support of the Engineering Society of Western Pennsylvania, an organization whose members included all of the city's leading industrialists, resulted in a vigorous debate over smoke abatement within the society and finally, several months later, in an official endorsement of a very narrow set of regulations to limit the smoke emitted by commercial boilers. Though this enabled the LHPA to win enough support in the city council for passage, in 1892, of a very weak, difficult-to-enforce smoke ordinance, the city's smoke problems continued to worsen.[14]

Undeterred, the women continued lobbying for a stronger, more enforceable law requiring smoke abatement from the factory furnaces that processed iron, steel, coke, and other materials used in manufacturing. They continued to reach out to men for help, and, as environmental conditions in Pittsburgh worsened, the men began responding more positively. In 1894, several prominent businessmen joined them in suing companies that were in clear violation of the 1892 ordinance for creating public nuisances, and several newspapers began publishing articles and editorials in favor of a stronger regulation. In May 1895, the city council bowed to the pressure by passing a new smoke ordinance that was slightly stronger than the 1892 law in some ways, though still quite weak. Opponents immediately began fighting in the courts to turn back the clock to the earlier, even weaker 1892 ordinance.[15]

Coalition building moved to a new level in October 1895, when the indefatigable women of the LHPA joined forces with reform-minded businessmen and women from other women's organizations to form the Civic Club, a mixed gender organization co-led by men and women that was committed to pursuing a broad range of municipal reforms. The Civic Club made smoke regulation one of its top priorities,

---

[14]Angela Gugliotta, "'Hell with the Lid Taken Off:' A Cultural History of Air Pollution – Pittsburgh" (Ph.D. dissertation, Department of History, Notre Dame, 2004), 183–231; John O'Connor Jr, "The History of the Smoke Nuisance and of Smoke Abatement in Pittsburgh," *Industrial World* March 24, 1913: 353–4; Stradling, *Smokestacks and Progressives*, 42–3, 207 n. 12. See also Angela Gugliotta, "Class, Gender, and Coal Smoke: Gender Ideology and Environmental Injustice in Pittsburgh, 1868–1914," *Environmental History* 5, no. 2 (April 2000): 165–93.

[15]Gugliotta, "Hell with the Lid Taken Off," 232–8; Robert Dale Grinder, "From Insurgency to Efficiency: The Smoke Abatement Campaign in Pittsburgh Before World War I," *Western Pennsylvania Historical Magazine* 61 (July 1978): 189–90.

but little progress was made until Andrew Carnegie, the city's best known steel maker, became a convert to the anti-smoke cause. In a well-publicized speech to the city's Chamber of Commerce in 1899, Carnegie decried the terrible impact the smoke nuisance was having on the city and called for the publically subsidized construction of manufactured gas factories outside the city (near the coal fields) and a pipeline system to distribute the gas all over the city, so that rich and poor, industry and residents alike could use gas instead of coal to heat and power their homes and factories.[16]

The solution Carnegie laid out was so inspiring that the Chamber of Commerce formed a Committee on Smoke Abatement and began working closely with the Civic Club on a solution to the smoke problem. Though the chamber leadership quickly realized that Carnegie's grand vision of a manufactured gas solution to the problem was far too costly, they pressed on with the broader goal of cleaning up the city's smoke through reform of the smoke ordinance. Working with the Civic Club, they lobbied strenuously for enactment of a rigorous new smoke ordinance with real enforcement teeth that would cover mills and factories as well as commercial buildings. In 1902 a court overturned the 1895 smoke ordinance, which forced the city to revert to the even weaker 1892 regulations, making the smoke problem even worse. The anti-smoke forces stepped up their campaign. In 1906, the newly established and reformist *Pittsburgh Sun* and other city newspapers entered the fray. With the help of front-page reports and editorials on the cost of smoke and the need for strong regulation, the anti-smoke coalition finally succeeded in forcing the city council to enact a strong and enforceable smoke ordinance that covered Pittsburgh's manufacturing as well as its commercial districts.[17]

The new ordinance sparked another intense backlash from manufacturers. The resulting litigation led to a state court decision in 1911 that declared the new regulations an illegal abuse of the city's police power. This compelled the city to revert back, yet again, to the weak 1892 ordinance. The anti-smoke forces immediately began a new campaign to persuade the state legislature to pass a bill to officially authorize Pittsburgh to regulate smoke. They won enactment of an enabling law a few months later. The city then passed a new ordinance that provided for professional inspectors and other much needed enforcement mechanisms. Unfortunately, however, the smoke opponents were not able to replicate their earlier, very hard won 1906 success in its entirety. Caving in to pressure from angry manufacturers, the city council exempted Pittsburgh's iron and steel mills from the new smoke ordinance's strictures, and their smoke continued to plague the city.[18]

The battle continued. Andrew and Richard B. Mellon, scions of one of Pittsburgh's premier industrial and banking families, founded a research institute

---

[16]Gugliotta, "Hell with the Lid Taken Off," 238–44.

[17]Grinder, "From Insurgency to Efficiency," 190–9; Gugliotta, "Hell with the Lid Taken Off," 247–313; O'Connor Jr, "History of the Smoke Nuisance and of Smoke Abatement in Pittsburgh," 354.

[18]Gugliotta, "Hell with the Lid Taken Off," 313–7; O'Connor Jr, "History of the Smoke Nuisance and of Smoke Abatement in Pittsburgh," 354–5.

at the University of Pittsburgh the year the new smoke ordinance was enacted and commissioned it to study the city's smoke problems. Publication, in 1913, of the first of the Mellon Institute's reports on the harmful impacts and high economic costs of smoke inspired the Pittsburgh Chamber of Commerce to convene a meeting of the city's business and women's organizations that led to the formation of a new anti-smoke organization, the mixed gender (though male-dominated) Smoke and Dust Abatement League. The league lobbied for passage of a stronger smoke ordinance and succeeded in securing passage of such a law less than one year later. With this victory in hand, its members pressed city officials to enforce the tough new regulations stringently. They also initiated campaigns to educate the public to the harm caused by smoke and the many economic and public health benefits of smoke abatement, as well as the availability of various kinds of equipment for reducing smoke emissions. These efforts were so successful that the number of Pittsburgh's "Smoky Days," as defined by the U.S. Weather Bureau, declined by 50% between 1912 and 1917. Of course, the 50% reduction in smoke was a far cry from complete elimination, but this was still a significant improvement. The reductions took place not only in the central business district, but also in industrial parts of the city that were not covered by the smoke ordinance. In fact, so many of the city's factories voluntarily reduced their smoke emissions (to save fuel costs) that it became obvious that the opponent's claim that it was economically and technically impossible for industry to abate factory smoke was wrong, and the city council expanded the smoke ordinance to cover the manufacturing district.[19]

After these successes, much of the energy seems to have gone out of the popular movement for smoke regulation in Pittsburgh. However, it resumed in the late 1930s, after the city eliminated its Bureau of Smoke Regulation. Again women reformers collaborated with businessmen to advocate for strong, new regulations. Among the groups involved were the Chamber of Commerce and the women-led Civic Club and League of Women Voters. They succeeded in enacting a new ordinance in 1941. It was modeled on the one enacted in St. Louis and regulated residential as well as industrial smoke. After its passage, two new groups, the United Smoke Council, composed of 80 allied Pittsburgh and Allegheny County organizations, and the Allegheny Conference on Community Development, a group committed to revitalizing the central business district and the regional economy, assumed leadership of the movement, working to help residents and businesses come into compliance with the new regulatory requirements. They also lobbied the state for the power to regulate railroad smoke. The city received this authority in 1947.[20]

---

[19]Gugliotta, "Hell with the Lid Taken Off," 368–450.

[20]Ibid., 476–614; Sherie R. Mershon and Joel A. Tarr, "Strategies for Clean Air: The Pittsburgh and Allegheny County Smoke Control Movements, 1940–1960," in *Devastation and Renewal: An Environmental History of Pittsburgh and Its Region*, ed. Joel A. Tarr (Pittsburgh: University of Pittsburgh Press, 2003), 145–73.

Learning from History

What lessons should today's community sustainability activists draw from the history of business involvement in smoke regulation in the late-nineteenth- and early-twentieth-century cities? I would argue that these histories represent a counter-argument to the lessons learned from the more recent wars over environmental regulation, during the 1970s, 1980s, and 1990s, when business firms and their trade organizations were the enemies, not the leaders, of pollution and other kinds of environmental regulation. While those more recent experiences taught environmentalists to fear and loath business, the history of smoke regulation demonstrates the utility of building coalitions with progressive business leaders to achieve strategic political and regulatory goals.

The time seems ripe for such collaboration again. Indeed, it is already starting to happen – at the national and international, as well as the local, community levels. Growing numbers of major industrial and financial corporations have become members or corporate partners of climate change projects sponsored by national and international environmental NGOs, like CERES, WWF, the PEW Center on Global Climate Change, the Aspen Institute, and the World Business Council for Sustainable Development.[21]

Some NGOs (and corporations) have also developed programs to restructure markets in ways that further environmental goals. The U.S. Green Building Council, for example, has promulgated a set of green building standards and awards that is driving the change not only in the building supply industry, but also in the architecture profession, as architects to bone up on green building design. The Forest Stewardship Council has developed a set of highly regarded sustainability standards and a certification process that is having a similar impact in the forest management and products fields.[22] CERES sponsors the Investor Network on Climate Risk (INCR). The over 60 institutional investors, investment banks, and other financial institutions that participate are leveraging their collective power to deploy their portfolio investments and shareholder rights to pressure Fortune 500 corporations to improve their climate policies, investment strategies, and carbon emission disclosure practices. In the absence of a U.S. government–sponsored carbon cap and trade market, the Chicago Climate Exchange (CCX) created a voluntary market that enables participating companies to make voluntary but legally binding commitment to meet annual GHG emission reduction targets, sell or bank CCX offsets if

---

[21] http://www.ceres.org/page.aspx?pid=705; http://www.worldwildlife.org/what/index.html; http://www.pewclimate.org/business; http://www.aspeninstitute.org/policy-work/energy-environment/our-program; http://www.wbcsd.org/templates/TemplateWBCSD5/layout.asp? MenuID=1. For a general overview, see Andres R. Edwards, *The Sustainability Revolution: Portrait of a Paradigm Shift* (Gabriola Island, BC: New Society Publishers, 2005), 49–112.

[22] http://www.usgbc.org/; http://www.fscus.org/

they reduce their emissions below the target level, or buy offsets if they miss the target.[23]

Much like the business-led smoke abatement organizations of old times, these organizations and programs provide technical advice and guidance to participating firms to help them reduce their GHG emissions and identify market opportunities to provide climate-protecting products and services to others. They also convene discussions with U.S. and international policy makers, organize conferences on policy developments, and encourage participating firms to lobby government officials and policy makers in favor of climate change regulations, while issuing reports designed to influence policy debates.

In addition to these high-profile, national and international programs, organizations, programs, and networks are proliferating at the regional, state, and local levels that encourage, provide technical guidance to, and otherwise work to develop environments for firms to improve their carbon disclosure, reduce their GHG emissions, develop and market environmentally friendly products, and make other changes needed for communities to become more environmentally sustainable. Groups of people in small businesses and professions are coming together to form sustainable business community groups as a way for their members to gain moral support and business connections for green business endeavors.[24] Trade organizations have committees and working groups that convene sessions at trade conferences that develop and promulgate voluntary standards and other guidance to help firms in their supply chains work with each other more cost effectively to reduce the environmental footprints of their products. They also advise state and local government authorities on sustainability issues, often in collaboration with local environmental NGOs.[25]

If the history of smoke regulation is any indication, these sustainability-oriented business organizations can do more than provide technical advice to community policy makers regarding environmental best practice. They can, like the anti-smoke businessmen of old times, also join forces with community environmentalists to lobby for the regulatory policies needed to achieve sustainability goals, leveraging their resources and political influence to help environmentalists pass the legislation and revenue measures needed to enable towns and cities to reduce their harmful

---

[23] http://www.incr.com/NETCOMMUNITY/Page.aspx?pid=198&srcid=-2; The Carbon Disclosure Project is a similar London-based international NGO. http://www.cdproject.net/; http://www.chicagoclimatex.com/

[24] http://www.sustainablebiz.org/; http://www.bayareaalliance.org/business-diff-sustainable.html. For links to a set of similar organizations in other states, see http://www.living economies.org/networks.

[25] Several semiconductor trade organizations have programs in this area. See, for example, http://www.sia-online.org/backgrounders_wsc_eshtf.cfm and http://wps2a.semi.org/wps/portal/_pagr/113/_pa.113/798. For activity at a local level in Silicon Valley, see http://svlg.net/campaigns/cleanandgreen/index.php. See also http://www.piba.org/sponsored_events.html (these events change over time).

environmental impacts and become more sustainable. If history is a guide, environmentalists would be wise to do as smoke reformers did and reach out to them with an eye toward working together to achieve shared goals.

The Darker Side of Business's Role in Smoke Regulation

Of course, environmentalists face risks working with business to achieve environmental goals. History also provides perspective on this darker side of the issue of environmental collaboration. Some historians have looked at business's role in the anti-smoke movement with a particularly critical eye. In their view, women reformers were the heroes of the movement, while businessmen and engineers were a relatively conservative force who tended to hold the movement back by focusing excessively on developing relatively low cost technical solutions to the smoke problem, rather than on simply cleaning the air. Businessmen and the other male members of the anti-smoke movement were, they claim, less willing to enact tough regulation than the women – and less prosecutorial in their approach toward enforcing such regulation.[26]

Perhaps because they view business's role in this negative way, smoke movement historians Harold L. Platt and David Stradling devote a great deal of time to the story of a particularly dark episode in the history of smoke reform that raises questions about whether environmentalists can trust their allies in the business community. Rather than a successful collaboration between business and non-business reform groups, this is a story of political betrayal and regulatory failure. I will recount it here because it illustrates, with particularly graphic force, the downside risk for environmentalists of working with the business community to achieve sustainability goals.

This particular episode took place in Chicago in the context of a crusade to solve the problem of railroad smoke. Chicago was a major railroad hub, filled with the tracks and terminal facilities of most of all the nation's many railroad companies. Heavy black smoke poured from the smoke stacks of steam-powered locomotives as well as the chimneys of depots, stations, and rail yards. In 1908, a group of South Side women declared war on the Illinois Central (IC)'s filthy smoke pollution

---

[26] Harold L. Platt, "Invisible Gases: Smoke, Gender, and the Redefinition of Environmental Policy in Chicago, 1900–1920," *Planning Perspectives* 10 (January 1995): 67–97; Platt, *Shock Cities*, 468–941; Grinder, "The Battle for Clean Air"; Stradling, *Smokestacks and Progressives*. For critiques of this interpretation and more positive assessments of business's role, see: Uekoetter, *The Age of Smoke*, 20–42; Frank Uekoetter, "Divergent Responses to Identical Problems: Businessmen and the Smoke Nuisance in Germany and the United States, 1880–1917," *Business History Review* 73 (Winter 1999): 641–76; Christine Meisner Rosen, "Business Leadership in the Movement to Regulate Industrial Air Pollution in Late Nineteenth and Early Twentieth Century America," *Economic History Yearbook* (2009 forthcoming). See also Rosen, "Businessmen Against Pollution in Late Nineteenth Century Chicago." These works focus on smoke reform during the Progressive Era. Mershon and Tarr, "Devastation and Renewal," 169–73, discuss the limits of the business–government collaboration that guided enforcement of smoke regulation in the 1950s and 1960s.

in 1908 with attention-getting parades, petitions, and threats that they would stop cleaning their homes and washing their children if the city did not use its power to solve the problem. Platt calls their crusade the "revolt of the housewives."[27]

Like female smoke reformers elsewhere, the South Side women quickly reached out to Chicago's male business community for help. They formed the Anti-Smoke League and quickly enlisted endorsements from what Platt calls a "diverse coalition of women's clubs, neighborhood improvement associations, and professional groups" (i.e., business organizations). The women and their anti-smoke business allies also developed a good working relationship with Mayor Fred Busse, a former coal dealer, and his Chief Smoke Inspector Paul Bird and his inspectors, who had been installed in city government as a result of previous business-led movements to regulate the smoke nuisance. Bird ordered a crackdown on IC for violating the smoke ordinance. As the movement gained steam, the city council met to consider legislation to force it to abate its smoke. The IC responded by agreeing to burn cleaner burning, but very expensive, hard coal.[28]

With this victory in hand, the city council began debating a proposal to force all the railroads to abate their smoke. They soon turned their attention to the possibility of requiring them to do this through the electrification of their tracks and passenger and freight terminals. The high cost of hard coal, coupled with the success of railroad electrification in New York City, made this solution to the problem attractive to the smoke reformers. Although the huge upfront costs of electrification made this mode of smoke abatement extremely expensive and even more controversial than the original plan to force the railroads to burn hard coal, the reformers saw it as a way to modernize and rationalize the city's rail system, while solving its smoke problem much more thoroughly than a switch to burning hard coal ever could.[29]

In 1909, the Chicago Association of Commerce (CAC), a business-led reform group that supported the smoke regulation movement, stepped into the fray with a seemingly helpful offer. The CAC offered to undertake a study of the smoke problem to help the supporters of smoke reform make the positive case for electrification with documentary evidence of its benefits and technical feasibility. Such studies, like the reports of the Chicago Citizens' Association made on the terrible condition of the Chicago River in the 1880s, were often undertaken by reform organizations to document the need for regulation and other government initiatives. The CAC's report, completed in 1910, concluded that electrification was both practical and economically feasible and recommended that it begin immediately.[30]

It was at this point that CAC leaders double crossed their allies in the anti-smoke movement. Before making their pro-electrification report public, they decided to run

---

[27] Platt, *Shock Cities*, 468–73; Stradling, *Smokestacks and Progressives*, 119–20.

[28] Platt, *Shock Cities*, 469–70, 485–6; Stradling, *Smokestacks and Progressives*, 120.

[29] Platt, *Shock Cities*, 485–7. See also David Stradling and Joel A. Tarr, "Environmental Activism, Locomotive Smoke, and the Corporate Response: The Case of the Pennsylvania Railroad and Chicago Smoke Control," *Business History Review* 73 (Winter 1999): 690–1.

[30] Platt, *Shock Cities*, 486–8; Stradling, *Smokestacks and Progressives*, 125–7.

it by a number of railroad officials, who persuaded them to quash it. The railroads saw little or no economic or managerial advantage to spending a fortune to electrify hundreds of miles of track as well as their extensive depot facilities and locomotives, simply to reduce smoke. Having lost the battle against electrification in New York City, and concerned that the U.S. Senate was beginning to look into the feasibility of forcing them to electrify the railroads entering Washington D.C., they were determined to do everything in their power to forestall electrification in Chicago. To do so, they persuaded the CAC leaders to begin a second, more "scientific" study, one that the railroads would generously fund, to make the case against electrification.[31]

The leaders of the CAC did more than agree to the new study. While keeping their original report secret, they persuaded the city council to postpone its vote on electrification until the new study was completed so that council members could make their decision on the basis of the new committee's well-researched evidence of the costs and benefits of electrification. Word of the suppressed 1910 report did not leak out for over 2 years. When it did, it generated considerable outrage, but by then, it was too late to save the movement.[32]

The new committee studied the issue for 4 years – two more than originally planned. The women leaders of the Anti-Smoke League and their allies continued to press for electrification, but as the years dragged on and the city council waited for the CAC's report, their movement lost momentum, while the railroads, through their intermediary, the Association of Commerce, gained control over how the smoke problem and its possible solution would be defined. For them it was a question of economics, not public health or civic beauty. The new report was finally published in 1915. Weighing in at an impressive 1177 pages and loaded with charts, graphs, and tables that gave it a distinguished air of scientific objectivity, it concluded that electrification was neither economically necessary nor economically feasible. Taking note of the committee's well-researched conclusions, the Chicago city council refused to enact an ordinance to regulate the industry's smoke. The report became an important tool in the railroads' political strategy for staving off attempts by other city governments to force them to abate their smoke by electrifying their tracks. It helped them block efforts to force them to implement this solution to their pollution problems for decades.[33]

---

[31] Platt, *Shock Cities*, 487–9; Stradling, *Smokestacks and Progressives*, 115–8, 120–1, 123–6.

[32] Platt, *Shock Cities*, 486–7; Stradling, *Smokestacks and Progressives*, 126.

[33] Platt, *Shock Cities*, 486–9; Stradling, *Smokestacks and Progressives*, 126–30; Chicago Association of Commerce and Industry. Committee of Investigation on Smoke Abatement and Electrification of Railway Terminals, *Smoke Abatement and Electrification of Railway Terminals in Chicago* (Chicago: Rand McNally, 1915). For an in-depth analysis of this episode from the perspective of the management of the Pennsylvania Railroad, see Stradling and Tarr, "Environmental Activism, Locomotive Smoke, and the Corporate Response," 689–702.

History's Complex Lessons

The CAC's successful effort to prevent Chicago city council from implementing railroad electrification regulation is certainly sobering. Does it mean that history shows that working with the business community is a strategy that is doomed to failure? No – I would argue that this is much too simplistic a conclusion. The CAC's betrayal of the smoke reformers' trust exemplifies one of the worst kinds of betrayal environmentalists can encounter when they work with business to achieve political goals. There was, however, much more to business's involvement in smoke reform than this.

History's lessons for environmentalists are written in the totality of business's role in the smoke movement – in the positive as well as the negative aspects of their role. Yes, business interests continually fought regulation and often succeeded in weakening legislation and thwarting enforcement. Yes, the CAC's betrayal of the movement was terrible. But this was only half of the story. In city after city, reform-minded business leaders fought for regulation, alone as well as with women and other anti-smoke reformers, often over decades, with remarkable commitment and dedication. They provided crucially important political clout, organizational capability, thought leadership, and technical research and development in the battles to enact smoke regulations and enforce them on their fellow business interests. As a result of their efforts and those of the other groups with whom they worked, the smoky pall over the nation's major industrial cities slowly cleared.

A growing number of business leaders appear to be positioned to do the same for community sustainability today. The history of urban smoke regulation suggests that the battle for regulations to advance community sustainability will be long and hard and slow and difficult. But business leaders can help environmentalists move the cause forward, just as they did during the Progressive Era. The wide range of interest groups organizing today to advance the cause of sustainability around the world arguably constitutes a new progressive movement, a global progressive movement – much like the progressive movement that gave rise to the urban smoke abatement crusades. These groups include business firms and trade organizations. If the past is any indication, community environmentalists need to recognize and take advantage of this. They need to put their concerns about the downside risk aside (not blindly, but with awareness of the upside potential for progress) and form alliances with those in their business communities who share their interests and have the political clout and the technical expertise to help them achieve their goals.

Chapter 3
Los Angeles Community College District (LACCD)

Larry Eisenberg

L. Eisenberg (✉)
Facilities Planning and Development, Los Angeles Community College District, Los Angeles, CA, USA
e-mail: eisenblh@laccd.edu

W.W. Clark II (ed.), *Sustainable Communities*, DOI 10.1007/978-1-4419-0219-1_3,
© Springer Science+Business Media, LLC 2010

Setting the Policy

As of now, the Los Angeles Community College District (LACCD) is a national leader in the implementation and development of sustainable building and renewable energy technology. The colleges and satellites of the district feature numerous Leadership in Energy and Environmental Design (LEED) certified buildings, and more importantly 100% of the energy need will be met through the use of cutting-edge renewable energy generation and energy storage technology. The story of the transition from being an aging educational colossus to a gleaming example of the best that architecture and technology can offer follows.

With more than 220,000 students a year attending classes, the LACCD is the largest community college district in the world. The nine colleges and major satellites serve the 882 square mile district and 36 incorporated cities. The infrastructure at the nine colleges was largely built in the 1940s, 1950s, and 1960s. With the passage in California of Proposition 13 in 1969, funding to public entities was severely curtailed and community colleges were no exception. As a result, for nearly 30 years, despite a rapidly growing population, the LACCD colleges saw little investment in new buildings and hardly any investment in maintenance of existing buildings. Time and intense use took their toll and eroded the quality of the buildings to the point that potential students living in the district chose to go to surrounding community colleges that had been able to invest in and maintain their basic building stock.

In 2001, the Board of Trustees of the LACCD asked the district's voters to approve a general obligation bond issue that would end the more than 30-year drought in construction and a growing deferred maintenance backlog for the nine colleges of the district. The voters in the nine campus districts approved not only the initial bond proposal of $1.245 billion to make an investment in affordable higher education, but also a second bond issue of another billion dollars in early 2003 to create a combined construction, renovation, and restoration program of $2.2 billion. Subsequently, voters approved another $3.5 billion in 2008, creating a total program funding base of $5.7 billion.

However, the opportunity presented by the initial LACCD bond issue did not escape notice of the environmental community in Southern California. Working closely with the district's Board of Trustees, environmental advocates educated and cajoled to explain the importance of a sustainable building program. The message was simple: Not only would a sustainable building program have long-term benefit for the district and its students, but it would also serve as an educational laboratory that could move sustainability into the mainstream.

After numerous hearings, prolonged conversation, and careful study, the Board of Trustees adopted the LACCD sustainable building and renewable energy policy in 2002. The policy established by the board required that all buildings built under the new bond program would be built at least to the then new LEED "Certified" standard. The policy also set a goal of self-producing 25% of the district's energy needs with at least 10% utilizing photovoltaic cells. The result would be more than 40 new buildings built to LEED standards and a few megawatts of renewable power put in place.

In retrospect, this policy, although bold and courageous in 2002, was really quite modest. The board consciously set a modest policy since no other large organization had set such an ambitious sustainable policy, and there was genuine concern that this policy would create significant added cost to the district's nascent bond program. In conversation about setting the sustainable policy, the trustees had been told that there would be a significant cost premium if they chose to build at the LEED "Silver" level rather than at the lower LEED "Certified" level. They were told that the "certified" level would result in a 10% cost premium. They were also told that the technology was not really available to support a significant renewable energy commitment. The reality in both regards proved to be quite different.

Implementing the Policy

In the fall of 2003, the district hired a new executive director for facilities planning and development with a background in large capital programs and with knowledge of and commitment to sustainable building requirements. He was charged with implementing the board policies for sustainable building and renewable energy. In turn, the executive director hired several talented individuals with a background in these disciplines to create a core sustainability/energy team.

The result is that LACCD is currently undertaking the largest sustainable building program in the United States. The more than 500 projects include renovations, upgrades, modernizations, and, most exciting of all, more than 40 new "green" buildings, representing the best in environmentally sensitive building techniques. Utilizing $2.2 billion in voter-approved funds, the district is executing an extensive building program to address much-needed campus improvements and transform its nine community colleges into state-of-the-art educational resources for students and the community.

The district will also be building two new college centers, one of which will be an adaptive sustainable reuse of the historic Van de Kamp Bakery in the Atwater area of Los Angeles. The other satellite project will be the adaptive reuse of the former Firestone factory in the City of South Gate to create a state-of-the-art education center and sustainable technology institute.

Recognizing that a $2.2 billion program has the power to change the marketplace, the Board of Trustees created a requirement that each design team, in which we currently have 140 working members, would include a certified LEED professional. When the program began, there were perhaps a dozen in all of Southern California. Today, there are nearly 400 certified LEED professionals in the LA area. The training that these architects and engineers obtained to achieve the certified designation will, of course, benefit not only the LACCD program, but also every future project that they touch.

The educational process would be enhanced in the design of the sustainable buildings, the buildings themselves when complete as examples of sustainable choices, and the development of an integrated curriculum component devoted to sustainability. Students and faculty along with the community would be partners and learners in building sustainable and smart college campuses. After all, the purpose of higher education is to research, innovate, and educate.

The LACCD Renewable Energy Program

Energy use and supply has become a significant component of the LACCD capital program. Living in sunny but energy-short California leads one to naturally think of alternate energy sources such as photovoltaic and solar heating solutions.

The modest renewable energy policy established by the Board of Trustees in 2002 has blossomed into a far-reaching energy program working on the cutting edge of energy technology. This policy has given license to explore a broad range of energy technologies including all forms of photovoltaic energy, energy storage, fuel cells, wind power, hydrogen generation and use, nanotechnology-based and industrial scale battery systems, anaerobic digestion, solar thermal, thermal storage, and geothermal concepts.

Energy and power for the LACCD colleges came from several sources that are central gird oriented. The Los Angeles Department of Water and Power (LADWP) is a municipal utility (MUNI) owned by the public and with a mayoral-appointed Board of Commissioners. It has six campuses in its jurisdiction. The Southern California Edison (SCE) is an investor-owned utility (IOU) company that supplies the other three campuses.

LADWP supplies more than 22 million megawatt-hours of electricity a year for Los Angeles's 4 million residential and business customers. It is the largest MUNI in California. The department's model is time tested and simple: a reliable and increasingly diverse supply of power, coupled with stable rates that are among the most affordable in the nation. This combination has effectively fueled the growth of Los Angeles for more than a century.

To improve system reliability and to ensure that power supplies continue to meet the city's needs for the next 100 years, LADWP is spearheading an aggressive Renewable Power System energy program (20% by 2010) to enhance generation capacity, modernize transmission and distribution infrastructure, assure power quality, and identify cost saving, environmentally sensitive efficiencies.

For the LADWP to move ahead into the future, people must be educated and trained at all ages about energy, water, waste, land, transportation, and air, which make up the component parts of sustainability development. Edison, on the other hand, has implemented many renewable energy projects. And since it is governed by the State of California Public Utility Commission, it has raised rates in order specifically to fund solar/pv systems for homes, offices, and complexes including college campuses. In 2006, the rate increases started to generate $300 million per year, which is to continue for the next 11 years.

Working closely with LADWP and SCE, the LACCD has established a program that will meet the need to train people of all ages about energy, water, waste, land, transportation, and air, which make up the component parts of sustainable development. The core concept involved in this effort is to use the energy infrastructure that will be put in place at the LACCD colleges and satellites to teach the technology. This effort will be supplemented with stand-alone touch screen kiosks and enhanced classroom capability to allow delivery of a sophisticated curriculum.

The energy generation program is at the core of the sustainable infrastructures for the nine college campuses and consists of four strategies that are now being implemented:

(1) Efficient renewable energy central plants

   (a) Build sustainable central plants that produce and deliver hot water and chilled water to heat and cool all college buildings.
   (b) Build a comprehensive four-pipe distribution system throughout each college to deliver and return the hot and chilled water.
   (c) Utilize solar thermal and geothermal as the only energy source to drive the heating and cooling processes.
   (d) Install thermal storage (ice storage) to reduce the afternoon peak heat load.

(2) Demand management/energy conservation

   (a) Conduct an investment-grade energy audit of all buildings on all campuses.
   (b) Retrofit all energy-consuming elements in all interior and exterior spaces for maximum efficiency (e.g., lights, fans, and pumps).
   (c) Install state-of-the-art and new technologies in all buildings (e.g., occupancy sensors and harmonic reduction filters).
   (d) Install individual building and building subsystem metering and monitoring systems to determine building and system energy use.
   (e) Add insulation, low e-glass, window film, white roofs, green roofs, etc.

(3) 100% renewable energy at each campus

   (a) Solar – solid panels, thin film, and concentrator technology
   (b) Wind – urban building-scale wind turbines
   (c) Geothermal – ground source heat loop
   (d) Hydrogen gas – utilize locally generated electricity to electrolyze water to produce hydrogen gas
   (e) Hydrogen storage – Utilize low-pressure solid-state hydrogen storage system
   (f) Fuel cells – PEM and alkaline fuel cells with direct hydrogen feed
   (g) Microturbines burning pure hydrogen gas
   (h) Lithium ion batteries for large-scale energy storage
   (i) Flow batteries for redundant energy storage
   (j) Hydrogen fueling for hydrogen fuel cell vehicles

(4) Sustainable development curriculum

   (a) Build on different campuses with basic focus courses as certificated, licenses and degrees.

(b) Career opportunities and training for jobs, new companies, and advanced degrees.
(c) Collaborate with unions, private businesses, public, government, and non-profit sectors.
(d) Provide actual experiences on campus through building programs.
(e) Sustainable development curriculum: solar, wind, geothermal, hybrids, as well as new businesses, accounting, operations, and maintenance.

Paradigm Shift

The desire to explore alternative energy concepts and the realization that the LACCD bond program will be adding 50% more square footage to the existing college building base led to a new thought: the idea that there will not be enough resources to effectively heat, cool, maintain, and clean this significant expansion in space. At the same time, the district realized that an alternative energy program had the potential to supplement its energy demand and, in the process, lower utility bills. Carried to its logical conclusion, alternative energy supplies could be developed in a manner that would entirely offset the district's energy bill. As noted above, at $9 million per year, this is a significant portion of the district's annual operating budget, and if redirected, could pay for a large number of additional maintenance and custodial personnel.

This combined thought led to the paradigm shift: a comprehensive alternative energy program could be the source of badly needed operational funding to meet pressing needs. With the development of significant alternative energy resources combined with either a long-term funding strategy or a shorter term buyout strategy, the district could lower its energy bill in the near term or eliminate the energy bill altogether in the long term.

To implement this concept, the LACCD has adopted a four-part energy strategy as outlined above. The components of the energy strategy are as follows:

(1) Efficient renewable energy central plants

Given the age of the LACCD colleges' facilities, for the very large part, the building's heating and cooling needs were met by the use of traditional rooftop heating and cooling units. In many cases, buildings were not even equipped with cooling capability, making teaching in the hot months a tenuous proposition at best. Some of the colleges were equipped with central plants that served part of the college heating and cooling need with chilled water and hot water based on the need at any point in time. These central plants utilized traditional boiler and chiller technology and distributed their chilled and hot water through a distribution network of pipes.

The first component of the new strategy is to move comprehensively to central plant provision of hot and cold water to all buildings. New central plants will include highly efficient boilers and chillers and will have the majority of their energy supplied through the use of a new technology – the Sun Chiller solar vacuum heat tube. These tubes are placed on the roof of the central plant, water is introduced into the tubes, and the sun heats the water to nearly the temperature of steam. The very hot water travels into the central plant, and through thermodynamics and heat exchangers, the energy is removed from the water and used to drive the chillers and boilers.

Similarly, for those colleges with an existing central plant, designs have been developed that provide for a smaller new central plant to supplement the existing central plant, creating a virtual central plant that relies on sophisticated energy management system software that blends their capabilities.

(2) Minimize energy demand through performance contracting

An effective energy strategy requires that each building be retrofitted to present the lowest energy demand possible. To make this occur, every single energy-consuming element needs to be analyzed to understand its potential to be retrofit with the most energy-efficient version available within an economically viable payback period, usually 7 years. In addition, the building needs to be analyzed for what does not exist, such as occupancy sensors, insulation, and multipane windows. A detailed building-by-building, room-by-room, device-by-device analysis needs to occur to determine what exactly needs to be retrofit and what needs to be added to the building in terms of state-of-the-art technology to allow it to present the absolutely lowest energy profile possible. Typically, a metering and monitoring system is added to evaluate the performance of the building over time and allow fine tuning of energy conservation measures to occur.

For the list of items noted in the report that have an economically viable payback period, there are companies that are willing to come in for free to retrofit and add the appropriate energy-conserving measures. These companies are paid back from the energy savings that they have guaranteed over a period of years. If the projected savings do not occur as predicted, the period of repayment lengthens to accommodate the amount of money available to pay back the initial investment. In any case, the recipient of the retrofits has no out-of-pocket expenses beyond its current cost of energy, which is now made up of two components: the amount used to pay back the performance contractor and the funds paid for utilities consumed.

LACCD is launching a comprehensive demand management performance contracting program to squeeze out every watt and therm possible from every building at each college. The result will be the minimal energy profile possible for every LACCD building. The cost will be borne by the performance contractor and supplemented through incentives and rebates available from the local utility companies.

(3) 100% renewable energy and energy storage

The last major physical component of the LACCD Energy Strategic Plan is the installation of photovoltaic generating capacity at each college and satellite facility.

One megawatt of rigid-frame photovoltaic panels using current technology covers approximately three acres and would cost approximately $7 million if purchased in today's marketplace on a stand-alone basis. Installed photovoltaics will cover the daytime peak load and provide a good amount of excess capacity. Three acres of panels requires the construction of carport structures over surface parking lots or the use of numerous building rooftops.

At a cost of $7 million per megawatt, photovoltaic panels would not be a cost-effective solution to meeting ongoing energy generation requirements. However, we are at a unique point in history where the combination of incentives and tax law has driven the cost of photovoltaic installations into the cost-effective range when done by a contractor that can utilize the tax incentives and rebates.

Cost factors that can be taken into consideration by a third-party financier that has the ability to use or sell tax credits include the following:

(1) the current federal energy tax credit
(2) the rapid depreciation capability of the internal revenue code
(3) availability of state and local rebates on solar projects
(4) the sale of the renewable energy credits (green tags) created with the installation of photovoltaic panels
(5) price reduction through bulk procurement or captive production concepts
(6) other federal investment tax credits

When used in combination, it is possible that costs of 10–20 cents on the dollar can be achieved for a photovoltaic array, bringing project payback well within normal standards for investment payback.

The other significant issue when considering a photovoltaic or wind-driven energy solution, for that matter, is the fact that the sun only shines during the day and wind does not typically blow in a consistent manner. If one is to provide solar or wind energy at night, or during the doldrums, there must be an energy storage technology employed. A host of new energy storage technologies are now being released that move far beyond the lead-acid battery and have storage capabilities in the multi-megawatt range. One example is a system that uses the excess solar power generated during the day and on weekends to electrolyze water and captures the hydrogen gas that is generated. The hydrogen gas is used as the fuel stock in a proton exchange membrane fuel cell and generates electricity and heat. The only byproduct in this system is water from the recombination of the hydrogen gas and oxygen in the fuel cell. The heat can be used to supplement central plant requirements or meet other site-specific needs such as heating a swimming pool.

A comprehensive photovoltaic/storage system will meet the needs of the LACCD colleges on a 24 × 7 basis with a redundant capability. The systems will be financed through the employment of a power purchase agreement where every watt generated and used by the colleges is paid for on a per-unit basis. The per-unit cost covers the cost of maintenance on the system and the cost of capital that pays back the final cost remaining on the system after the employment of all available tax credits, incentives, and rebates. The initial agreement developed by LACCD has established a per-watt charge of 13 cents compared to the present and escalating cost from SCE of 16 cents per watt. This difference results in an instant reduction in electricity cost to the college.

The LACCD power purchase agreements will contain a buyout clause that will allow LACCD to purchase the system at any time, but in particular after year 5 (end of the rapid depreciation period), to eliminate the per-watt charge. At that point, LACCD will own the means of energy production and eliminate its energy bill.

(4) Sustainable development curriculum

The final component of the LACCD Strategic Energy Plan is the incorporation of the scientific, business, and environmental lessons learned into the curriculum of the colleges. The concept is that our students across all disciplines can benefit from sustainable curriculum elements in either an academic or a vocational context.

The successful execution of the LACCD Strategic Energy Plan has the real potential to take the LACCD colleges off the grid and presents an organizational model that can be readily replicated by any organization. The model offers not only an environmentally sound energy strategy but a comprehensive budget strategy that will allow organizations to supplement operational resources to meet other high-priority needs.

Sustainable Practices

The LACCD commitment to sustainability extends to campuses being "smart" in other areas of the extensive program as well, including the purchase of furniture and concrete. We have established a zero landfill policy that will reuse or recycle all items that come from our construction activity. A central feature of this program is to avoid sending mountains of 30-, 40-, and 50-year-old furniture to the dump. With a little work, the district determined that old college furniture; equipment, and modular buildings could see a second life in the hands of countless nonprofit organizations in Southern California. Through partnerships with local organizations, our reuse program will not only save us money in terms of avoided transportation and landfill fees, but also provide a direct value to those who voted for our bond program.

For the replacement furniture and equipment, thanks to the support of the progressive national furniture industry, we will be buying high-quality 100% recyclable furniture that comes with a fully unlimited 15-year warranty at the best prices available. This was possible only because the major manufacturers were willing to change their factories to meet our large purchasing requirements. The use of our huge purchasing power in support of our sustainable goals has changed the way products are produced for the benefit of all.

Interestingly, myths remain about the wisdom of building green. A recent survey sponsored by Turner Construction reported continuing doubt about the costs of building green. The reality is that the huge volume of construction that is making the green choice, as our program did, has changed the marketplace. Green products are readily available at comparable prices. We have found that it is just a matter of making the right choice at the beginning and finding design professionals who will support sustainable goals. The LACCD program is a showcase for what is possible, but it is clear that higher education building programs across the country are getting the message loud and clear.

Water Conservation

Even though 70% of our earth's surface is covered with H_2O, there are many parts of the world – especially developing countries – where it is in short supply. Even for us living in Southern California, water is a precious resource that we must conserve in order to ensure that we have enough to meet our daily needs. The LACCD recognizes this fact and recently began a process to install 1,224 cartridge-type and non-cartridge type waterless urinals at all nine community colleges.

Waterless urinals are important to the district's sustainability program because they will benefit the district in three ways:

a. Each urinal will save approximately 40,000 gallons of water per year. Altogether that's a saving of almost 50 million gallons of water a year (that's enough water to fill 259 Olympic-sized pools!);
b. Less water consumption will result in a decrease in water bills;
c. Waterless urinals will eliminate the need to send wastewater to treatment plants, therefore reducing the sewage costs across the district.

Purchasing Carpet

1. Specifications reduced the number of ounces per square yard of carpet while retaining performance with a 30-year warranty. This reduces manufacturing waste, production cost, transportation cost, and energy used in the production process.
2. Specifications required that the LACCD carpet be made from solution-dyed fibers. This choice avoids the use of 50 gallons of water per yard of carpet if it were to be dyed after production. The after-production dying process requires that the water be heated to boiling. Solution dying avoids the need to boil this large quantity of water and saves large amounts of energy, reducing the generation of greenhouse gases.
3. The LACCD contract calls for very low tolerance for VOC emissions. This improves the safety of the occupants and helps broader environmental air quality.
4. The carpet contract is a "closed loop" contract. It requires the use of recycled content in the face and backing of the carpet. It also requires old carpet to be taken back for 100% recycling.

5. The contract specifies carpet tiles for most areas. The use of carpet tiles allows
   for efficient use of the material, eliminating much of the installation waste.

Concrete

The problem is that these "impervious" surfaces, such as parking lots, send pollutants directly into our waterways and ultimately into the ocean. According to the
United States Environmental Protection Agency (EPA), as much as 90% of these
pollutants, such as oil and antifreeze, run directly off the surface of traditional impervious surfaces into our rivers, streams, and oceans during rainstorms. With the need
to balance development with preservation and to minimize its impact on the environment, the LACCD has looked to other alternatives for its construction projects.
One such alternative has been the use of pervious concrete so that stormwater seeps
into the ground.

Pervious concrete is a mix of coarse aggregate, cement, water, and little or no
sand. Although used in some areas for decades, this mixture is generating renewed
interest. Due to its open-cell structure, which allows rainwater to filter down to
underlying soil, it is an excellent choice for mitigating pollutants and reducing the
quantity of storm water runoff when used in parking areas, low-traffic streets, plazas,
and walkways. Because of its practicality and effectiveness, this concrete actually replenishes aquifers, conserves and protects water supplies, and is American
Disabilities Act (ADA)-friendly.

As a result of its lighter color and lower density, it also helps to enhance air
quality by lowering atmospheric heating and decreasing heat island effects. (The
heat island effect occurs when tree-covered areas are replaced with dark pavement
surfaces. It is a contributing factor to the fact that an urban area can be up to 12°
hotter than its surrounding countryside).

In addition, pervious concrete's lighter color naturally reflects heat and light,
which studies have shown can save as much as 30% in lighting costs over other
pavements.

Using pervious pavements can also reduce the need for large wet pond detention
and retention systems allowing for more effective land use in addition to decreasing costs for labor and construction and maintenance of detention ponds and other
stormwater management systems. Furthermore, it decreases or eliminates the need
for expensive irrigation systems.

Pervious concrete is also quite durable. Areas properly designed and constructed
will last 20–40 years with little to no maintenance. As a result, it is widely
recognized as the lowest life cycle cost option (an analysis of the savings generated throughout the average life of a product compared to other similar products)
available for paving.

Construction Waste

A heavy construction phase of our program has begun. That means old and temporary buildings are being razed to make room for more efficient and technologically

advanced structures. Classroom and offices are being renovated and upgraded. Parking lots are becoming parking structures or new buildings. But while these wonderful improvements are what our program is all about, each of them creates construction waste.

Construction waste is one of the largest contributors to landfill volume in the United States. Our region's landfills are already in high demand. Communities are advocating closing existing sites and strongly protesting the creation of new locations. As it becomes increasingly expensive to bury our waste, governments are seeking alternative solutions. Construction waste is a major target of waste reduction efforts because it is a major contributor to the need for more landfills, requires an extensive transportation system to haul materials, and is a tremendous waste of the resources that have great recycling and use potential.

Finding secondary markets (users of diverted waste for new purposes) is both environmentally and economically rewarding.

Base rock, used for roads and building pads, is expensive. However, recycled concrete and asphalt, crushed on-site, is a viable alternative to purchasing new base rock and at a fraction of the price, not to mention the cost and pollution benefits of eliminating in- and outbound dump trucks.

As classroom buildings are renovated, many materials have potential for secondary markets. Wood, metal, plastic, glass, cardboard, sheetrock, and carpet, found in outdated classroom buildings, can all be recycled. In fact, all of the classroom and office furniture offered through LACCD's Furniture Value Program is made, in part, with recycled content.

By seeking out local secondary markets for our construction waste, we are supporting the local economy, extending the life of our existing landfills, having a significant positive effect on the environment, and of course making good financial sense.

LACCD is in the process of implementing our goal to better process our construction waste. We are looking into on-site sorting of basic materials into bins and then having them transported, hopefully at no cost, to the appropriate secondary market. The general contractor on each site would be tasked with implementing the sorting of materials for their respective work site, and bins would be picked up by the secondary-market vendor when full.

Inevitably, after the wood, carpet, metal, sheetrock, concrete, and asphalt have been removed, there will be some leftover "trash." This bin would be sent to a more sophisticated sorting facility for further reduction, which would divert even more waste away from the landfill. Our goal is to divert at least 90% of our construction waste away from landfills and into productive uses. Our ultimate, if ambitious, goal is to divert it all and send no construction waste at all to local landfills!

Resources are scarce, as all of us who have to deal with the cost of cement and steel are acutely aware. I strongly believe that our advocacy of construction waste recycling will help to change the market we rely on. By providing our "waste" to the local secondary market, we will provide the raw materials that will create the future products we will need to purchase and save us the cost of the dumping fees. If other contractors and construction sites join us in our support of construction waste recycling, the cost of sustainable materials will continue to decrease. By working together we can all benefit together.

Training the 21st-Century Workforce

A recent analysis of the California Air Resources Board (CARB) proposed a plan to cut carbon dioxide emissions to 1990 levels by 2020 estimates that California's productivity would potentially increase by $27 billion over what would be realized if the state didn't make the cuts. The "greening" of California could also mean 100,000 more jobs in 12 years and an increase in per capita income of $200 annually.

As the housing and financial markets continue to suffer, many local governments are looking to green industries and technology as the "future" for economic success and vitality. As those industries continue to grow in response to the industry's evolution and overall need, there is a greater necessity for trained workers that are in step with the technology and prepared to take that technology to the next level.

There are more than 18 million students at 6,500 colleges and universities around the country. While many colleges are focused on greening their campus, a study by the National Wildlife Federation found that only 53% of the more than 1,000 campuses examined offered environmental studies majors or minors. This is particularly true of disciplines that aren't traditionally connected to the environment or environmental issues. While overall growth may be slow, the good news is that there are more than 600 environmentally focused institutes at schools throughout the country.

Beyond promoting a more sustainable future, institutes for higher learning would also do well to consider presenting initiatives, education, and training programs in an environment that further promote environmental awareness and that directly and indirectly impact students in a positive way:

- Colleges designed with proper ventilation, materials selection, and acoustical quality have been found to improve student and employee health, which leads to improved attendance figures;
- Attention to site planning and adequate daylighting has been shown to heighten student performance by as much as 25%;
- Energy and water operating costs can be reduced by 20–40% through the use of new technologies and conservation methods;
- When advanced technology and design are made visible, buildings can become teaching tools and important features of college curriculum.

Educating more than 220,000 students per year, the LACCD is in an ideal position to not only provide our region's green workforce with the education and job training they need to succeed, but also lead by example on global issues that can make a positive impact on our environment.

In 2008, the LACCD announced the development of a new curriculum, which will provide both advanced technical and basic skills training to unemployed and underemployed residents in our region to improve their job placement. The programs and services are sector specific and target high-wage, high-demand industries such as sustainable energy/utilities, technology, and construction to name a few. After completing their studies, students will be able to obtain jobs in the full range of career paths in each of these sectors.

But learning doesn't just take place in the classroom or lecture hall. That is why LACCD extends its education efforts about sustainability to the community as well. Once a month, the LACCD's boardroom doubles as a lecture hall as attendees gather to listen to notable experts on every important issue of our environment, both local and international, learn about the problem, and become an active part of the solution. This sustainability collaborative is free and serves as a networking activity where attendees can learn about all aspects of a sustainable future.

Conclusion

The LACCD is aware of its role as both an institution of higher education and a role model for the community, which is why it has embarked on an ambitious campaign to raise awareness on the benefits of sustainability. On a state-wide level, its Board of Trustees has actively participated in the League of California Community Colleges, sharing sustainable building policy implementation techniques with other boards of trustees; has participated in UC/CSU sustainability conferences, Green Build Conferences, and the state-wide partnership between energy companies and colleges; and has addressed many neighboring school district boards to encourage duplicative efforts in other school districts.

On an international level, the board has been a leader in the field of sustainable development by becoming the first in the world to seek both LEED^{TM} and England's Building Research Establishment Environmental Assessment Method (BREEAM) credentials. LACCD acquired the historic Van de Kamp Bakery building, which will become the first construction conversion project in the United States to achieve a BREEAM rating. BREEAM seeks to minimize the adverse effects of new buildings on the environment while promoting healthy indoor conditions for the occupants.

Seventeenth-century British author John Donne said, "No man is an island, entire of itself; every man is a piece of the continent." The LACCD wholeheartedly agrees, which is why, over its long history, the district is proud of the various partnerships and collaborations it has established. During the development of the district's sustainability policy, board members turned to leading environmental groups to brainstorm and draft what would become the district's award-winning green policy. The partnership provided not only a good source of intellectual capital and experience, but also strong leadership and support when it counted.

If we are to act as leaders for the next generation, we should understand that it's important for us to prepare for the future, educate those that will take the reins, and lead by example. That is why the LACCD is embarking on a sustainable path of growth and development and, at the same time, imparting what we learn to the students who will become the business, government, and community leaders for the next generation.

Many people already know that the LACCD's $5.7 billion building upgrade and renovation program is one of the largest sustainable (that is, "green") building programs in the United States. But the sustainable mission of the LACCD is greater than simply designing and constructing buildings certified by the US Green Building Council. The LACCD view is much more holistic and includes supplying classrooms with sustainably built furniture, seeking out the most energy-efficient appliances, and developing a method to develop 100% of our own energy needs. As the mission and goals are implemented, a keen eye is being kept on the long-term cost efficiency.

A core principle of the LACCD Bond construction program, which is updating nine community colleges to better serve its students, is to incorporate best practices in sustainable design, construction, and operations wherever it can. With each sustainable step, the LACCD is taking us one step closer to our goal of balancing the need for development with protecting our environment for generations to come.

Chapter 4
Sustainable Cities as Communities and Villages

Patrick Stoner

P. Stoner (✉)
Program Director, Local Government Commission, Sacramento, CA, USA
e-mail: pstoner@lgc.org

W.W. Clark II (ed.), *Sustainable Communities*, DOI 10.1007/978-1-4419-0219-1_4,
© Springer Science+Business Media, LLC 2010

Fig. 4.1 A home in Davis, CA with passive and active solar systems

A growing number of cities have been "going green." While this concept has different meanings in different communities, it reflects a growing concern of large and small cities to be aware and proactive about their environment, their resources, and their impact on climate. Now organizations throughout the United States and around the world have grown into powerful groups, advocating a range of sustainable strategies. The Local Government Commission (LGC) is one such organization.

The LGC is a nonprofit, nonpartisan membership organization that provides inspiration, technical assistance, and networking to local elected officials and other dedicated community leaders who are working to create healthy, walkable, and resource-efficient communities. The LGC's membership is composed of local elected officials, city and county staff, planners, architects, and community leaders who are committed to making their communities more livable, prosperous, and resource efficient.

Much of LGC's work since the early 1990s has been guided by the Ahwahnee Principles for Resource-Efficient Communities,[1] which the LGC developed with the assistance of noted architects and planners. The Ahwahnee Principles focus on the following:

- Integrated communities that mix uses essential to the daily life of residents;
- Community size that allows easy walking for daily needs and with a central focus;
- Transit, and pedestrian and bike paths that connect to all destinations;

---

[1] The Ahwahnee Principles can be found at http://www.lgc.org/ahwahnee/index.html.

- Diverse housing types for a variety of economic levels and ages;
- Ample open space whose frequent use is encouraged by placement and design;
- Well-defined edges to communities or clusters of communities that are protected from development;
- Preservation of natural terrain, drainage, and vegetation;
- Resource conservation and waste minimization;
- Street and building orientation that improve energy efficiency; and
- Use of local materials and methods of construction.

The LGC has worked primarily with local governments in California for the past 30 years, and so much of this information comes from there.

In the early 1990s, the US DOE provided funding to Santa Monica and San Jose, California, and to Portland, Oregon, to develop sustainable city programs. All three adopted a program in 1994, and they continue to be leaders in the sustainability arena. Today if one googles "Sustainable Cities," 645,000 responses are found. This is testament to the commitment level of government closest to the people to be a steward of the environment and local economy.

In California, the State has mandated waste diversion and energy efficiency measures that require local government actions toward sustainability. Adopted in the 1980s, AB 939 required cities and counties to divert 50% of the waste going to landfills in 1990 by the year 2000. Aggressive recycling, composting, and reuse programs were the result, and most of California's over 500 local governments met the diversion goal. California's Title 24, which provides an energy usage standard for new residential and commercial construction, has been and continues to be stricter than federal standards or any other state's building codes. The State's energy efficiency strategic plan now has a goal of zero net energy use by 2020 for new residences and by 2030 for new nonresidential construction. Title 24 and California's history of energy efficiency programs have kept per capita electricity consumption in the state stable, while it has risen in the rest of the country.

California has a new impetus toward sustainability, the Global Warming Solutions Act of 2006 (AB 32). It provides ambitious goals to reduce greenhouse gas emissions statewide by 2020 and 2050. California's attorney general has sued cities and counties that have not addressed climate change when they updated their general plans. While local government actions toward meeting the statewide goals are at this point voluntary, many cities and counties view AB 32 as a wake-up call to action if they have not already initiated a climate action plan or as ammunition to help them implement their plans if they have.

California's Assembly Bill 117 allows cities and counties to aggregate the electric load of their residents, businesses, and institutions to provide them electricity. The majority of local governments investigating community choice aggregation (CCA) have been interested in at least doubling the amount of renewable generation that the State requires of the utilities currently serving their constituents, and they see CCA as an economic (hedging against future higher cost, fossil fuel–based electricity), environmental (reducing greenhouse gas emissions from the electric sector), and sustainability strategy.

These California mandates make it both easier and harder for cities and counties to be "sustainable." It's easier because the statewide bar is set so high that some degree of sustainability is required. It's harder for the same reason; surpassing the state standards means going after some of the higher hanging fruit. Many cities and counties have taken steps in particular areas such as smart growth, renewable energy, energy efficiency, recycling, and stormwater management. A few of them are highlighted below.

Emeryville, California

The City of Emeryville, California, is a small community (less than 2 square miles) located near the eastern end of the San Francisco Bay Bridge, which links San Francisco to Oakland and Berkeley. It is a very dense community with high-rise residential buildings and large commercial centers. There is very little of Emeryville that is not covered with buildings or pavement.

Emeryville has worked for a long time on reclaiming, remediating, and redeveloping the many brownfields within its borders. These efforts sparked a successful economic rebound. The city did not stop there and decided to harness redevelopment for better environmental outcomes, in particular that related to stormwater runoff. The city faced several challenges, including a high water table, clay soils, and few absorbent natural areas among the existing and redeveloped industrial sites.

In December 2005, the Emeryville City Council adopted *Stormwater Guidelines for Green, Dense Redevelopment: Stormwater Quality Solutions for the City of Emeryville*. These guidelines outline ideas for meeting new stormwater treatment requirements using site design, parking strategies, and stormwater treatment measures to allow water to flow through plants and soil. Numeric requirements apply to projects of 10,000 square feet or more. The guidelines generally require vegetative stormwater treatment measures and apply citywide.

The guidelines include a spreadsheet model for sizing on-site BMPs. The main strategies fall into several categories:

- Reduction in the parking footprint by way of shared parking, making the best use of on-street parking and pricing strategies;
- Landscape design features, such as tree preservation and planting, use of structural soils, and bioretention and biofiltration strategies;
- Water storage and harvesting through cisterns and rooftop containers; and
- Other strategies to handle or infiltrate water on development and redevelopment sites.

Alameda County, California

StopWaste.Org is the Alameda County Waste Management Authority and the Alameda County Source Reduction and Recycling Board operating as one public agency. It manages a long-range program for development of solid waste facilities and offers a variety of other programs in the areas of source reduction and recycling, recycled product procurement, market development, technical assistance, and public education. The voters of Alameda County imposed a per-ton disposal fee on themselves that has put StopWaste.Org in a unique position to go beyond the usual waste management projects of most local governments.

StopWaste.Org has been able to provide technical assistance and services related to green building, procurement, landscaping, and climate protection to communities in the county. While these topics are related to waste reduction (regionally appropriate landscaping can reduce the amount of green waste going to a landfill, for example), few communities have the budget to fund such exercises. Much of StopWaste.Org's work has been borrowed and adapted by other California communities.

Green building strives to improve design and construction practices so the buildings built today will last longer, cost less to operate, and contribute to increased productivity and better working environments for workers or residents. It is also about protecting natural resources and improving the built environment so that ecosystems, people, enterprises, and communities can thrive and prosper. Green building represents a paradigm shift – a crucial change in the way we understand, design, and construct buildings today. It doesn't happen by accident; it requires thorough planning, thoughtful design, and quality construction.

To achieve these goals, green building promotes a whole-systems approach to the planning, design, construction, and operation of buildings. This comprehensive approach benefits communities, residents, and businesses by:

- Improving construction quality;
- Increasing building longevity;
- Reducing utility, maintenance, and infrastructure costs;
- Protecting the health of workers and residents; and
- Enhancing quality of life in our communities.

StopWaste.Org has a wealth of resources for remodeling and new construction of single-family homes. These include design and construction guidelines and directories of green building products and services. There are also comprehensive Multifamily Green Building Guidelines to address the unique needs and requirements of multifamily housing.

StopWaste.Org has developed an Environmentally Preferable Purchasing Model Policy to consider how a product is made and what it's made with. Beyond recycled content, it is also important to look at other environmental attributes of products, such as energy consumption, toxicity, air and water pollution impact, materials efficiency (such as packaging), and the disposal impact at the end of its useful life.

StopWaste.Org's Bay-Friendly Landscaping program views the built landscape through an environmental lens, in order to improve the health of the community and watershed. The principles and practices of Bay-Friendly Landscaping provide a framework to help "green" the built landscape. For public agencies, Bay-Friendly means that civic landscapes can model practices that:

- Provide a sense of place and are suited to the local climate, soils, and topography;
- Reduce waste and help meet recycling goals;
- Reduce water use on landscapes by 50% or more;
- Prevent or reduce stormwater pollution to local creeks and the bay;
- Lower maintenance associated with mowing and shearing; and
- Reduce greenhouse gas emissions.

Bay-Friendly Landscaping staff work with the Alameda County Waste Management Authority's 17 member agencies to help make informed decisions about sustainable landscaping in their communities. Bay-Friendly resources currently available for public agencies in Alameda County include landscape guidelines, grants and design assistance, model policies and ordinances, workshops, scholarships, and custom trainings.

Bay-Friendly Landscaping is a whole-systems approach to the design, construction, and maintenance of the landscape in order to support the integrity of one of California's most magnificent ecosystems, the San Francisco Bay Watershed. The Bay-Friendly Landscape professional can create and maintain healthy, beautiful, and vibrant landscapes by:

- Landscaping in harmony with the natural conditions of the San Francisco Bay Watershed;
- Reducing waste and recycling materials;
- Nurturing healthy soils while reducing fertilizer use;
- Conserving water, energy, and topsoil;
- Using integrated pest management to minimize chemical use;
- Reducing stormwater runoff; and
- Creating wildlife habitat.

A well-designed and maintained Bay-Friendly Landscape can cost less to maintain in the long run by consuming fewer resources. For public spaces, Bay-Friendly Landscapes embody community values for health, wildlife, and the environment. For private property, Bay-Friendly Landscapes addresses issues that your clients care about, such as lower water or garbage bills as well as increased environmental benefits.

StopWaste.Org is also supporting the efforts of Alameda County cities to conduct municipal and community greenhouse gas emission inventories, adopt reduction targets, and develop a model climate action plan for the county.

Marin County, California

Marin County lies north of the Golden Gate Bridge from San Francisco. Like much of the San Francisco Bay area, it is progressive in areas related to protecting the environment. Marin has several sustainability efforts underway to help residents, businesses, and institutions reduce their impact on the environment. The county has a partnership with Pacific Gas and Electric Company (PG&E) to provide energy efficiency programs and services and a plan to create a new green power agency to provide renewable electricity to its constituents.

Marin County Energy Watch is the county–PG&E energy efficiency partnership. Energy Watch provides energy management services to public sector agencies, a small business energy assistance program, building tune-up services to make sure the energy systems are operating properly, training for community youth to provide a variety of free energy-efficient and water-saving devices in qualifying residences, and an energy wise program for realtors.

The county is also working to free itself from overreliance on nonrenewable energy with its Marin Clean Energy (MCE) program. MCE is looking to take advantage of California's CCA law, which allows cities or counties, or groups of them, to provide electricity to their constituents. The law (AB 117), passed in 2002, covers the electrons only; the transmission and distribution of the electricity remains with the existing utility, in Marin's case, PG&E. MCE would involve a joint power authority between the county and some, if not all, of the cities within the county.

While other communities investigating CCA have focused on providing electricity at a cost less than existing utility rates, Marin has focused on maximizing the amount of renewable energy generated and consumed in the county. Studies have shown that providing 100% renewable energy would cost up to 10% more than PG&E rates (which by state law will include 20% renewable energy by 2011). For customers not wanting to pay more for electricity, MCE expects, due to the financing advantage of municipal versus private entities, to be able to provide a higher percentage of renewable energy at the same price as PG&E. Surveys done of Marin residents have shown enthusiastic support for higher renewable percentages, even if it will cost more.

The promoters of MCE view this renewable strategy as a hedge against future price increases in fossil-fueled generation. Does anyone expect the cost of petroleum to drop as supplies become scarcer? The expectation is that eventually this near-100% renewable generation strategy will remove the communities in the county from the uncertainty surrounding foreign oil. Support is growing among city and county elected officials, but there is still much to do before the MCE will become a reality.

Berkeley, California

The City of Berkeley has long been a leader in sustainability. Along with San Francisco, Berkeley adopted time of sale energy efficiency improvement ordinances. The Residential Energy Conservation Ordinance (RECO) and Commercial Energy Conservation Ordinance (CECO) undergo periodic review to ensure that they do not put an onerous burden on buyers and sellers of property in the city.

The ordinances kick in at the time of sale or if renovations to the building equal $50,000 or more. Information about the programs can be found at: www.ci.berkeley. ca.us/SubUnitHome.aspx?id=15404.

Berkeley also pioneered the Sustainable Energy Financing District idea in California. The Berkeley Financing Initiative for Renewable and Solar Technology (FIRST) aims to make it easier for residents and businesses to install solar energy systems and energy efficiency improvements on their properties. With current state and federal subsidies, installation of solar electric and solar thermal systems is cost effective for many residential and commercial property owners. However, disincentives to installation remain, particularly the high upfront cost and other financial hurdles.

The citywide voluntary Sustainable Energy Financing District allows property owners (residential and commercial) to install solar systems and make energy efficiency improvements to their buildings and pay for the cost as a 20-year assessment on their property tax bills. No property owner would pay an assessment unless he or she had work done on his/her property as part of the program. Those who do have work done on their property pay only for the cost of their project and fees to administer the program. The city would secure the upfront funding through the placement of a taxable bond. The financing mechanism is loosely based on existing "underground utility districts" where the city serves as the financing agent for a neighborhood when utility poles and wires are moved underground.

The Sustainable Energy Financing District solves many of the financial hurdles facing property owners. First, there are few upfront costs to the property owner. Second, the upfront costs are repaid through a voluntary tax on the property; therefore, funding approval is not determined directly by the property owner's credit or equity in the property. Third, the total cost of the solar system and energy improvements is comparable to financing through a traditional equity line or mortgage refinancing because the well-secured bond will provide lower interest rates than are commercially available. Fourth, the tax assessment is transferable between owners. Therefore, if the property is sold prior to the end of the 20-year repayment period, the next owner takes over the assessment as part of the property tax bill.

The Sustainable Energy Financing District was developed as part of the city's implementation of Measure G – the ballot measure setting greenhouse gas reduction targets for Berkeley.

Davis, California

The City of Davis, in California's Central Valley, traces its sustainability roots back more than 40 years when it decided to establish the first bike lanes in the United States. Other early actions that included conserving energy, producing solar energy, protecting farmland and habitat, and adopting innovative land use policies have helped establish Davis as one of the early leaders in the sustainable communities movement. The city is one of the many communities across the nation that has acknowledged its role in reducing climate change impacts that are threatening the planet. Davis is also home to Village Homes, an innovative development designed to encourage conservation of energy and natural resources and building a sense of community.

Davis started by painting lines on streets to provide validation and a place for bicyclists in the community, and currently it has over 100 miles of bike lanes and paths that meander throughout the city and connect every neighborhood with downtown, commercial centers, and each other. Davis was the first of only two cities in the United States to be designated at the platinum level for being a bicycle-friendly community by the League of American Bicyclists. It was also the first city to have dedicated bike turn signal lights at major intersections.

Walking, buses, and trains are also promoted in Davis. The city and the University of California at Davis cofunded Unitrans, the bus system that carries university students and other residents throughout the city. And intercity rail riders boarding in Davis consistently rank in the top three in the number of users out of all the communities on the route that includes Sacramento, Oakland, and San Jose. This diversified transportation system will figure heavily in the city's effort to reduce greenhouse gas emissions.

The City of Davis has many programs to protect natural resources. Its Open Space Program includes planning, land acquisition, site development, and management. It has developed an open space acquisition and management plan, a habitat conservation plan, a farmland preservation ordinance that conserves agricultural land and requires land use buffers as mitigation for new urban development land conversion, a wildlife habitat acquisition program, and dual use of the city's stormwater detention system as wildlife habitat and open space areas. The 400-acre Davis Wetlands Project is part of a growing effort to preserve and restore native habitats and the wildlife they support.

The city's Urban Forestry Program began in 1963 when it established its Street Tree Committee. Before that time, developers were encouraged to plant one tree in the area between the road and the sidewalk for each home they built, in order to provide shade and esthetics. The city now provides a master tree list of appropriate species and locations and a tree planting and maintenance program. Davis is one of the several Central Valley communities that require tree planting and maintenance in new or remodeled parking lots in order to provide shade, reduce urban heat island effects, and reduce the evaporation of volatile organic compounds and other noxious fumes emanating from parked cars.

The Davis Community Garden is sponsored by the city and operated through the Community Services Department. There are over 100 garden plots, for which the city provides land, water, and some maintenance and administrative support. The Davis Farmers' Market was started in 1975 and is one of the most well known and successful in the state and serves as a center of local community life and culture. The Davis Food Co-op started in 1972 and is now currently owned and operated by over 9,000 local households.

Davis is focusing on addressing local greenhouse gas emissions by conducting an inventory and setting reduction targets. It is coordinating with the university and

Fig. 4.2 Natural drainage system at Village Homes in Davis, CA

other organizations to identify where efforts on sustainability and climate change overlap. It has joined the US Conference of Mayors' Climate Protection Agreement and the Cities for Climate Protection Program. The city has set up a Climate Action Team to develop a Greenhouse Gas Reduction Plan for the community.

Village Homes is a 70-acre subdivision located in the western part of Davis that was designed to encourage the development of a sense of community and conserve energy and other natural resources. Its construction started in 1975 and continued through the 1980s.

Village Homes' streets are narrow (less than 25-feet wide) cul-de-sacs that connect with the bike and pedestrian trails running throughout the project. They are oriented east–west so that the homes all have south-facing walls to take advantage of passive solar energy and are ready to accept active solar water-heating and photovoltaic systems on the roofs. The narrow street design made them easier to shade, reducing the urban heat island effect for this neighborhood in the hot Central Valley of California.

Homes are generally oriented away from the streets, which seem more like alleys, and toward common green spaces that separate home lots and contain pedestrian/bike paths and the natural drainage system for the community. The drainage system includes a network of creek beds, swales, and pond areas that allow rainwater to be absorbed into the ground rather than being carried away in storm drains.

A total of 40% of the land in the Village Homes subdivision is commonly owned, including the greenbelts, orchards, and two large parks. More than 30 varieties of fruit trees and a vineyard form a large part of Village Homes' edible landscaping.

The combination of design features in Village Homes promotes outdoor activity that allows neighbors to regularly interact and build a sense of community. And the total package has been good for home prices in Villages Homes, which have consistently been at the high end per square foot for homes in Davis.

Web Site Links

Alameda County StopWaste.Org: http://www.stopwaste.org/home/index.asp
Berkeley FIRST: http://www.ci.berkeley.ca.us/Mayor//GHG/SEFD-summary.htm
Davis Sustainability: http://www.cityofdavis.org/cmo/Sustainability/
Emeryville Stormwater Guidelines: http://www.ci.emeryville.ca.us/planning/stormwater.html
Local Government Commission: www.lgc.org
Marin Clean Energy: http://marincleanenergy.info/
Village Homes: http://www.villagehomesdavis.org/

Chapter 5
Greening Existing Communities

Christine S.E. Magar

Introduction

In his book *The Future of Life*,[1] E.O. Wilson points out that if the earth's 6 billion-plus inhabitants were to increase consumption rates to the level we maintain in the

C.S.E. Magar (✉)
AIA, LEED AP, MED
e-mail: cmagar@greenform.net

Illustrations by Robert Clark

[1] Wilson, EO. *The Future of Life*. New York: Knopf, 2002.

W.W. Clark II (ed.), *Sustainable Communities*, DOI 10.1007/978-1-4419-0219-1_5,
© Springer Science+Business Media, LLC 2010

United States, we would require *four* planets to meet the world's demand for natural resources. Clearly, generating additional planets is not an option. Still, unsustainable development (and consumption) continues apace throughout the world, in the context of a scientific consensus that climate change effected in part by human activity (increased carbon dioxide emissions associated with deforestation, fuel combustion, and other sources) poses grave consequences for human civilization.

In the United States, the scientific consensus on global warming has yet to result in meaningful governmental legislation or regulation to limit carbon emissions by industries or communities. However, the scientific consensus has produced a substantial *popular* consensus, which has spurred individual and communal action to address concerns about the impact of unsustainable human activity on climate and finite environmental resources.

For architects, engineers, environmentalists, planners, and other key players and stakeholders, working toward sustainability via new green buildings, hybrids, and recycling programs is fairly straightforward. But most communities are not blank slates: effective sustainability strategies must account for existing buildings, transportation systems, energy infrastructure, and industry.

This chapter shows how one community is overcoming the challenges associated with greening its existing buildings and infrastructure via a business improvement district planning process. Significant to this approach is its treatment of the community's buildings as an interrelated whole – it is a given fact that sustainability cannot be achieved by concentrating on one building or system at a time.

The planning process seeks synergies across buildings, land use types, utilities, and other shared services to achieve sustainable solutions that save money (and even generate revenue), while they increase energy efficiency, reduce carbon emissions, lower demand on natural resources, and improve quality of life for members of the community.

WCBID Commitment

The Wilshire Center Business Improvement District (WCBID) was created by the Los Angeles City Council in 1995 and is one of more than 30 business improvement districts in the city. WCBID's area covers 100 acres, and more than 100,000 people live, work, or shop in its 33 million square feet of residential, office, and retail space.

Members of WCBID include merchants, institutions, and property owners who together set goals and map strategies to:

> Help Wilshire Center become a more livable and workable area, to provide for a better overall environment both socially and economically, to work towards constructive change for Wilshire Center, and ... advocate[e] ... in the area of public safety, beautification, maintenance, and economic development ...[2]

---

[2] *WCBID Annual Membership Report*, available at http://www.wilshirecenter.com/work/wcbic.htm.

Against the background of a significant popular consensus on global warming, and in the context of the organizational mission described above, in 2007 WCBID made a commitment to reduce carbon emissions by 80% by 2050, initiating the first Cool District in North America.[3] To achieve this goal, the board of directors voted to reduce carbon emissions by 2% a year for 40 years.

The WCBID pledge is consistent with state and local efforts to address climate change, including the California Global Warming Solutions Act of 2006,[4] which seeks to reduce greenhouse gas emissions to 1990 levels by 2020, and the *Green LA: An Action Plan to Lead the Nation in Fighting Global Warming*,[5] a City of Los Angeles undertaking with the goal of reducing greenhouse gas emissions to 35% *below* 1990 levels by 2030.

From Pledge to Vision

Pledging to reduce carbon output is a necessary first step, but it remains a first step. WCBID itself did not know how to achieve its stated goals. Buildings alone are responsible for 40% of greenhouse gas emissions. A public event for Earth Day 2007 brought the district's commitment to the attention of the local American Institute of Architects (AIA) office. Once aware of WCBID's pledge, the Los Angeles AIA's Committee on the Environment (COTE/LA) offered to help transform it into a reality. COTE/LA provided the expertise needed by a community that was willing to act, but unsure how to proceed. It defined specific goals and a framework for the planning and implementation process.[6]

At the same time, an effort was initiated to establish WCBID's carbon footprint – without identifying a current baseline, it is impossible to measure whether reduction goals are being met. However, establishing a baseline turns out to be a tremendous challenge. There is no consensus on how – for example – vehicle miles should be measured. Should this include only vehicle miles driven within WCBID? Or should the baseline account for *total* miles driven by WCBID residents and workers?

This challenge of establishing a baseline is still being addressed by WCBID and its partners.

In 2008, the COTE/LA team enlisted WCBID members, building owners, retrofitting experts, engineers, landscape designers, transportation engineers, and local utility representatives in a series of brainstorming meetings, or eco-charrettes, to develop specific measures for addressing climate change in the community. The

---

[3]Cool Cities is an initiative established by the Sierra Club to provide a framework for local communities to take action on climate change. More information is available at http://coolcities.us.

[4]This law incorporates a timetable to bring the State of California in line with the provisions of the Kyoto Protocol. More information is available at http://gov.ca.gov/press-release/4111/.

[5]Available at http://www.lacity.org/ead/EADWeb-AQD/GreenLA_CAP_2007.pdf.

[6]The author led the COTE effort and established the framework for collaboration with WCBID.

interdependence of systems, organizations, structures, infrastructures, and the environment was highlighted in these discussions, and sustainability principles underlay all discussions.

The first eco-charrette focused on establishing a sustainable vision for the community and generating short-, medium-, and long-term measures to achieve that vision from the perspective of open space, infrastructure, and transportation. The second eco-charrette largely focused on carbon reduction opportunities associated with WCBID's predominant building types: high-rise commercial and low-rise residential structures. It should be noted, however, that both meetings addressed all categories to some extent, regardless of the putative focus. This reflected the interdependence of the different categories (e.g., building improvements may rely on infrastructure enhancements and vice versa).

The third eco-charrette, which is ongoing at the time of this writing, entails a series of targeted public meetings that bring together diverse WCBID stakeholders to discuss specific resource management issues. Experts deliver presentations on waste management, water conservation and recycling, energy efficiency and electricity generation, transportation, and open space and conduct discussions that address questions about how these resources could be more efficiently managed in the WCBID. The third eco-charrette serves multiple purposes, including educating stakeholders on key issues associated with carbon reduction, eliciting feedback, and generating grassroots commitment to carbon reduction goals.

Underlying the Vision

For effective problem solving, it is important to recognize that the WCBID, like any community, is a complex system. System parts are typically not as freestanding as they may appear. For example, a change to transportation infrastructure – a new subway stop – may have significant effects on nearby businesses, residences, and open space. With an established vision for the whole community, proposed solutions, strategies, and incentives tend to recognize the interrelatedness of community systems. Decision makers can better determine the effort, effectiveness, and likely impact of suggested changes. Also significant, a vision with clear underlying principles empowers any participant to be a decision maker: beneficial change is generated by individuals and groups who see opportunities to bring the vision closer to reality.

In the context of meeting carbon reduction goals, a key principle was to make aspirational decisions. That is, decisions must account for future and possibly nonobvious benefits. Such benefits include community benefits (e.g., from open space allowances, traffic reduction, urban farms), environmental benefits (e.g., increased carbon absorption from "green" roofs and street plantings), and long-term economic benefits (e.g., renewable energy generation, gray water reclamation).

Underlying discussions of solutions were a set of sustainability principles, formulated by William McDonough for EXPO 2000 at the World's Fair in Hannover, Germany. The Hannover Principles support a holistic vision for design, problem solving, and decision making that encourages the consideration of interdependencies and unintended consequences and rejects short-term thinking. The principles are summarized below:[7]

- Insist on rights of humanity and nature to coexist.
- Recognize interdependence.
- Respect relationships between spirit and matter.
- Accept responsibility for the consequences of design.
- Create safe objects of long-term value.
- Eliminate the concept of waste.
- Rely on natural energy flows.
- Understand the limitations of design.
- Seek constant improvement by the sharing of knowledge.

These principles also reflect a commitment to systems thinking and are especially useful for groups seeking to frame a vision for achieving long-term goals.

Eco-Charrette Process

The eco-charette process was developed by COTE/LA specifically for the collaboration with WCBID. The objective was to generate a sustainable and achievable vision for the community, as well as short-, medium-, and long-term measures to achieve that vision. The results of the first two charrettes were documented in the form of illustrated brochures that were provided to the WCBID community.

The first eco-charrette, which focused on generating a vision for a sustainable community, engaged approximately 40 regional experts, including architects, engineers, sustainability specialists, and industry leaders in green construction, urban design, transportation, water cycling design, and other areas. Attendees were organized into four small groups, each led by two COTE/LA representatives, to develop visions related to residential buildings, commercial buildings, transportation, and infrastructure.

The second eco-charrette had a more concrete focus. Two representative buildings in the community – a commercial high rise and a residential low rise – were selected as prototypes. In advance of the eco-charrette meeting, COTE/LA members toured the two buildings and obtained architectural drawings. Local utilities were engaged to perform electricity, natural gas, and water use audits of the two buildings.

The approximately 40 people attending the second eco-charrette included the owners of the prototype buildings, representatives of local utilities and city and state agencies (such as transit providers and wastewater management specialists), as well as sustainability specialists and COTE/LA members. Participants were organized

---

[7]McDonough, William. *The Hannover Principles: Design for Sustainability*. Available at http://www.mcdonough.com/principles.pdf.

into three small groups, which were tasked with developing short-term, medium-term, and long-term carbon reduction recommendations for residential buildings, commercial buildings, and transportation infrastructure.

For these two eco-charrettes, one individual in each small group was responsible for documenting ideas in terms that a layperson would understand, while another was responsible for developing representative illustrations. The brochures were developed by COTE/LA in concert with a writer and a graphic designer. Brochures have been distributed throughout the WCBID community; in addition, brochure content appears on the WCBID web site.

Envisioning a Sustainable Business District

Recommendations from the eco-charrettes are discussed below. The following goals were identified as key to achieving the sustainable vision:

- Act as single entity/community to take advantage of proximities and synergies.
- Reduce automobile dependence.
- Increase density.
- Create a stronger sense of community.
- Reduce energy consumption and increase local energy production.
- Reduce water consumption and stormwater runoff.

Group Action

To achieve some objectives, it is advantageous for the whole community to act in unison. The recommendations in this section highlight strategies that leverage group synergies. For group negotiations, a leader is required. WCBID plans to take the lead for the recommendations in this section, enrolling and enlisting community members and building owners as appropriate.

Short Term

- Negotiate utility contract savings for the entire business improvement district.
- Secure pledges from building owners to upgrade buildings and improve efficiencies.
- Secure pledges from building owners to install a percentage of photovoltaics and/or wind turbines for local energy generation.
- Negotiate reductions on retrofit materials (e.g., light bulbs, window films) for the entire business improvement district (this would entail securing pledges from building owners to purchase a certain amount of materials over a period of time).
- Initiate green leases[8] with building tenants.

---

[8] More information is available at http://www.sustainca.org.

Midterm

- Audit the district to determine existing synergies between current waste products and required resources. Connect building owners and business owners to trade and share resources and wastes.

Reduce Automobile Dependence

A significant share of greenhouse gas emissions comes from auto exhaust, so a key goal for WCBID is reducing the use of motor vehicles. WCBID consists of a dense collection of high-rise office buildings, large hotels, regional shopping complexes, churches, entertainment centers, and diverse residential buildings. The vision of a car-poor (if not car-free) community is feasible.

The key to reducing automobile dependence is going on a "road diet." This means reducing lanes of traffic and dedicating some road space for public transit, bicycles, pedestrians, and green space. Aside from reduced auto traffic, a road diet results in more vibrant retail environments and walkable streets.

Note that traffic-related recommendations entail negotiations with the Department of Transportation, City Council members, transit officials, and others.

Short Term

- Designate bus-only lanes on the primary boulevard – in this case, Wilshire Boulevard.
- Designate bike-only lanes on the primary boulevard.
- Close streets on a temporary basis as on Farmers' Market Days. (This requires finding interesting retail activity where such closures would create a pedestrian- and retail-friendly environment.)
- Provide bus passes and/or parking cashout (cash payments for not using parking spaces)[9] to commuters.
- Install bike rental stands and/or building bike fleets to reduce auto usage for daytime errands.
- Establish a bike-sharing program.
- Share parking space with neighboring entities such as temples, churches, and farmers' markets to encourage telecommuting and discourage automobile use.
- Increase parking near transit to attract riders.
- Reduce parking and encourage telecommuting by offering incentives.

---

[9]More information about parking cashout programs is available at http://www.sierraclub.org/sprawl/transportation/cashout.asp.

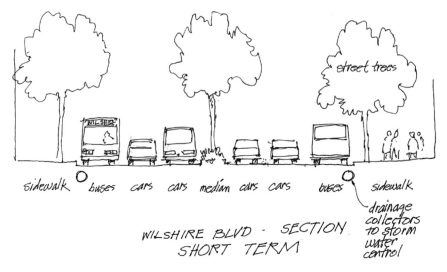

sidewalk   buses   cars   cars   median   cars   cars   buses   sidewalk

drainage
collectors
to storm
water
control

WILSHIRE BLVD · SECTION
SHORT TERM

Fig. 5.1 A better urban boulevard

Midterm

- Establish a Transportation Management Association. This organization would create literature, signage, and programs to locally promote carpooling, telecommuting, bicycling, and public transit.
- Minimize parking access for residential buildings.
- Cut street curbs adjacent to buildings to facilitate carpool drop-off.
- Add bike parking, bike racks, showers, and lockers to buildings.
- Expand the network of bicycle lanes connected to the city and county systems; increase availability of bicycle racks and lockers at transit stations.
- Employ traffic-calming road designs, including roundabouts on side streets and parking decks.
- Lobby for more bus and minibus routes through and within WCBID.
- Designate bus-only lanes on secondary avenues, in this case, Western Avenue and Vermont Avenue.
- Designate appropriate tertiary streets, in this case, 6th and 8th streets for through traffic.
- Designate appropriate tertiary streets, in this case 4th, 5th, and 7th streets as pedestrian/bicycle thoroughfares.
- Build area parking structures of several stories to free up surface-level parking for conversion to parks, retail, and residential uses.

Long Term

- Designate primary boulevard – the WCBID's "main street" – as car free, with lanes for buses and bicycles and a bio-swale median strip.

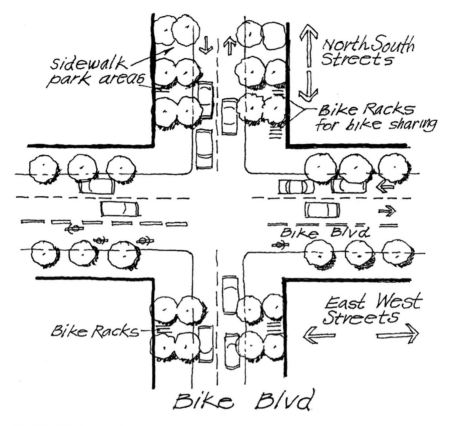

Fig. 5.2   Bike boulevard

- Establish modal investment of transportation dollars. This means setting percentage goals for travel by foot, bicycle, auto, and transit – and investing transportation funds accordingly.
- Transition parking lots entirely to nonauto usage as use of the single-user auto is drastically reduced.
- Decrease parking near transit to support transit-oriented development (TOD).
- Promote public–private partnerships to develop property and open space adjacent to transit stations.

Strengthen Community

It is a happy coincidence that many measures designed to reduce carbon footprint simultaneously enhance quality of life. Greening our environment – whether literally, with gardens that capture carbon, or figuratively, with bicycle lanes that

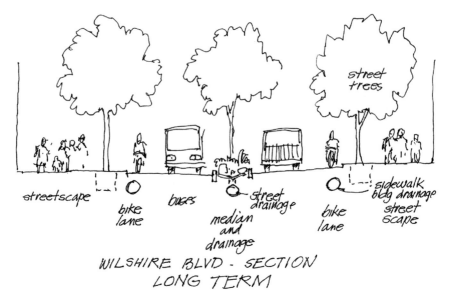

Fig. 5.3 The ideal urban boulevard

encourage human-powered travel – offers many opportunities for building stronger, healthier communities.

It is worth noting that moneys from city and state government programs can often be leveraged to support planting efforts. Because plantings reduce stormwater runoff, rebates or credits from utilities may also be available.

Short Term

- Create new streetscape plantings, including food-producing trees.
- Identify open space that can be converted to parks.
- Plant trees inside buildings, where they can serve as bioreactors that capture and sequester carbon before it gets in the atmosphere.
- Encourage volunteerism by offering opportunities within the community.
- Create green spaces in and around residential buildings, including pool areas.
- Develop a community garden or urban farm.

Midterm

- Designate 50% of north–south streets as green. Redirect them for one-way autocirculation and install new stormwater gardens.
- Identify open spaces for parks and recreation opportunities.
- Implement on-site composting in buildings.
- Implement training and demonstrations of sustainable behaviors.

Fig. 5.4 Exterior vines increase air quality and decrease heat gain

- Commission light-colored artwork for parking lot surfaces to reduce heat island effects and create an amenity.
- Create a community spaces with a building courtyard or a building rooftop.
- Build window boxes for each living unit for herbs and flowers (this will also reduce heat gain and improve air quality).

Long Term

- Establish rooftop gardens for insulation, stormwater reduction, and tenant amenity.
- Plant vines on building exterior walls to improve air quality and reduce heat gain.
- Integrate transit-oriented development with open space.
- Replace parking spaces with native landscaping, urban farm, and compost bins.
- Create and encourage cohousing, intentional communities, and eco-villages.
- Use publicly owned land to satisfy local needs for food, energy, water, and waste supply and management.

Conserve and Generate Energy

Energy from nonrenewable sources is a key contributor to greenhouse gas emissions. Transitioning to renewable energy sources can take time. Reducing consumption

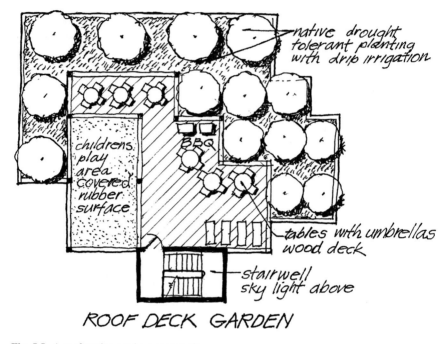

native drought
tolerant planting
with drip irrigation

childrens
play
area
covered
rubber
surface

BBQ

tables with umbrellas
wood deck

stairwell
sky light above

ROOF DECK GARDEN

Fig. 5.5 A roof garden can be a community space

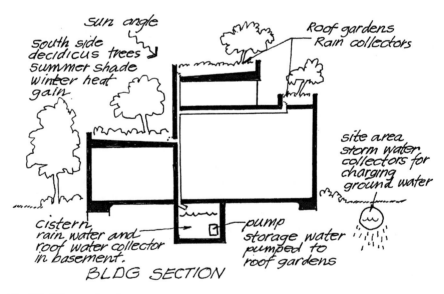

sun angle

south side
decidicus trees
summer shade
winter heat
gain

Roof gardens
Rain collectors

site area
storm water
collectors for
charging
ground water

cistern
rain water and
roof water collector
in basement.

pump
storage water
pumped to
roof gardens

BLDG SECTION

Fig. 5.6 Urban roof gardens

is equally important, and many conservation recommendations can be adopted immediately. Note that energy-related recommendations typically involve partnerships, negotiations, and incentive opportunities with local utilities and government agencies.

Short Term

- Upgrade to compact fluorescents or LEDs in residential units.
- Upgrade garage lighting to T8s with electronic ballasts.
- Install occupancy sensors.
- Install Energy Star-compliant ceiling fans.
- Install Energy Star-compliant HVAC units.
- Set hot-water temperature at lowest acceptable setting.
- Install window film on south- and west-facing glazing to reduce heat gain.
- Routinely maintain HVAC ducts, fans, and pumps.
- Conduct an energy audit and identify benchmark goals based on results.
- Conduct retrofit commissioning to ensure all systems are optimized.
- Evaluate existing heating system and determine whether replacement with a more energy-efficient system is feasible.
- Explore on-bill financing for energy efficiency upgrades.
- Locate and exploit energy synergies with other consumers.
- Continually investigate and leverage new energy technologies as they become available (e.g., solar, geothermal, hydrogen).
- Consider use of renewable energy sources at or adjacent to transit stations.
- Utilize rooftops, south sides of buildings, and parking lots for photovoltaic and wind-turbine systems.
- Leverage electric and gas utilities for rebate and incentive programs that are already in place.

Midterm

- Repair, seal, and insulate ducts. Shut off air supply to vacant spaces.
- Add insulation, and use a white coating when reroofing.
- Deactivate radiant heating and cooling, and replace with heat pumps for heating and cooling in residential buildings.
- Install solar thermal heating for swimming pools.
- Install electrical meters over phone lines in residential buildings.
- Design and install awnings at windows in residential buildings to reduce heat gain.
- Install variable air volume (VAV) HVAC systems. These systems enable building owners to control heating and cooling by zone. VAVs offer substantial energy savings for large buildings.
- Add utility- or building-owned photovoltaic systems.
- Plant street trees and solar panels in open space adjacent to buildings to reduce heat gain and generate electricity.

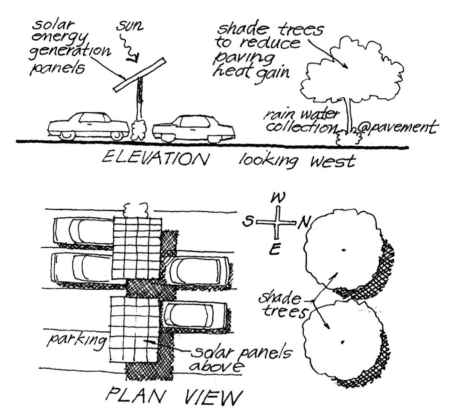

Fig. 5.7 Shading parking with solar panels and trees

- Install energy consumption meters in building lobbies, and encourage energy-reduction competitions among tenants. This also requires submeters for each floor.
- Rewire existing electrical systems and install photocell sensors to facilitate daylight harvesting.
- Utilize existing methane sources (by drilling existing surface parking lots or new green streets).
- Identify existing geothermal opportunities to produce local energy at district level.
- Utilize the metro tunnel under the primary boulevard to share excess power.
- For photovoltaic installations, take advantage of available city, state, and federal incentive programs.

Long Term

- Design and install light shelves and photosensors to reflect light into residential courtyards and facilitate daylight harvesting.
- Install skylights and tubular daylighting pipes on top floors of residential buildings.

- Replace single-glazed windows with dual-glazed windows.
- Install solar thermal hot water heating.
- Install photovoltaic panels to generate electricity from the sun.
- Convert parking lots to photovoltaic solar farms to generate the electricity to meet the needs of neighboring buildings. (Power generated could eventually be used for charging electric cars).

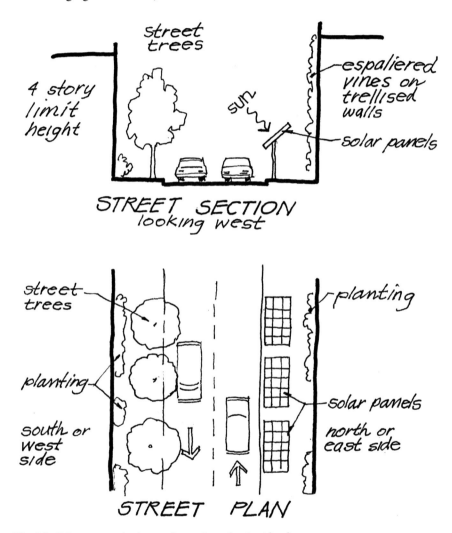

Fig. 5.8 Urban trees and solar panels coexist at the street level

- Create new building exterior frames to support photovoltaics, create new air space, and facilitate insulation and natural ventilation.
- Create a WCBID-wide solar strategy that leverages available city, state, and federal incentives.

- Develop a local, Central Energy System to meet all WCBID energy needs, based on PV, wind, fuel cells, geothermal energy, or biogas.
- Implement district heating/cooling plants (located in parking areas freed up by reductions in requirements).

Conserve Water

Rain Harvesting

It should be obvious that conserving water conserves energy. The various systems that process water, from drinking water supply to surface water management to wastewater treatment to heat water, all use a great deal of power. Reducing water consumption and capturing and reusing water before it enters stormwater and wastewater systems save significant energy – not to mention freshwater, which is a scarce resource.

Assistance and incentives are often available from local utilities and government agencies for water conservation measures.

Short Term

- Install low-flow aerators and/or flow restrictors on faucets.
- Install low-flow showerheads in residential units.
- Check for water leaks. Sudden changes in monthly billings may provide clues to leaks.
- Inspect and repair irrigation system monthly.
- Carry out water leak inspections annually.
- Replace plumbing fixtures with ultralow or no-water fixtures, such as waterless urinals.
- Reduce use of bottled water by adding a water filtration system.
- Replace existing landscaping with drought-tolerant species.
- Replace existing irrigation systems with low-water systems.
- Fill open, unused spaces with planters, and add trees to parking lot islands, to add shade and filter rainwater.
- Establish native plants, including native street trees, in all public spaces.
- Limit water used for cleaning in transit stations.

Midterm

- Provide education about water conservation, including reducing consumption and water supply, stormwater, and wastewater systems.
- Install ultralow flush toilets.
- Collect rainwater for landscaping in rain barrels or cisterns.

Fig. 5.9 Urban rainwater harvesting

- Install a septic tank.
- Acquire a fire sprinkler tank to store water to be used for irrigation.
- Reuse fire sprinkler water for irrigation of green roof gardens and street trees.
- Investigate and harvest nuisance water that collects at the base of buildings and reduces flow.
- Reconfigure the existing street fabric to include curb projections and green bus stops to filter water.
- Install plantings and landscaping (including food-producing trees) to filter stormwater.
- Add native plantings, including trees, in open space, such as parking lot islands.
- Reduce stormwater runoff from parking lots by installing permeable paving.
- Convert asphalt and concrete to decomposed granite (DG), or grass pave with native grasses and shade- and drought-tolerant plants to capture and infiltrate or reuse stormwater.
- Grade outside surfaces to direct stormwater to existing infiltration planters.
- In parks, plazas, and other open spaces, provide mechanisms for stormwater runoff collection.
- Design and build green screens (green screens are an architectural wire mesh installed on a facade to support vine growth) around parking structures.

Long Term

- Add storage tanks to harvest roof stormwater; use collected water for irrigation.
- Over open parking structures, add a photovoltaic canopy that can direct rainwater for capture.
- Replace sidewalks with pervious surfaces.
- Over closed parking structures, create parks or community gardens.
- Install bioswales in the primary and tertiary boulevard median to filter water.

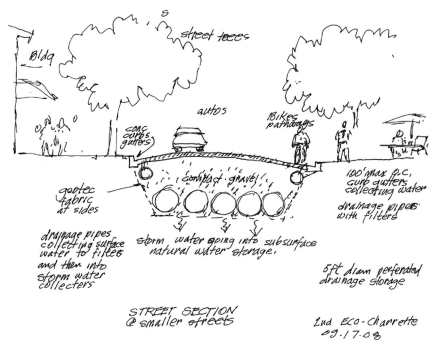

Fig. 5.10 The ideal urban street

- Install gray water system for irrigation and toilet flushing.

From Vision to Reality

The challenges posed by global climate change can seem formidable – even overwhelming – because the human contribution to greenhouse gas emissions is interwoven into almost every aspect of how we live, work, and play. The enormity of the problem can induce a sort of paralysis, even in those who appreciate the urgent need for action.

The example of WCBID's partnership with COTE/LA demonstrates one approach that communities can use to effectively address vast and complex issues. As we have seen, the first step entails a shared commitment to a common goal.

Second, it is necessary to create a shared vision. COTE was able to assist WCBID with the creation of a shared vision by engaging professionals who brought their diverse sustainability expertise to bear on the specific characteristics of the WCBID community. It was equally important to introduce sustainability principles that recognized the interdependencies between building, infrastructure, and transportation systems and weighed long-term environmental, economic, and community benefits against short-term costs.

The third step is communication. Through brochures, WCBID web site content, and meetings, the sustainable vision and associated recommendations have been made available to the wider community. The third series of eco-charrettes more directly engages members of the community in discussion about particular sustainability issues, such as energy conservation and transit.

The fourth step is action. Individuals and organizations can feel comfortable taking action because they understand where they fit in the shared vision of a sustainable community. They can easily determine the impact of their actions because the opportunities for short-, medium-, and long-term action have been clearly identified.

The WCBID endeavor has underlined the importance of partnerships. Most obviously, WCBID partnered with COTE to develop a sustainable vision and generate recommendations to achieve that vision. Partnerships with local utilities and government agencies that offer expertise and incentives for achieving sustainability goals are also vital. For example, WCBID members may leverage local utilities for energy audits, grants, and other services.

Clearly, WCBID is a work in progress. Baselines for greenhouse gas emissions are still being developed, and sustainability recommendations and strategies are being communicated and prioritized. Next steps for the community include the setting of specific targets and measurement of results. However, WCBID has succeeded in developing a framework for addressing climate change that is easily transferable to other communities; in fact, the work accomplished is likely to be mimicked by other communities, in Los Angeles and elsewhere.

> We're hoping to be a model, the greenest business improvement district in North America. The idea is to get other business improvement districts to take on this challenge. . .[10]

---

[10]Gary Russell, executive director of WCBID, in *New Angeles Monthly*, October 2008. Available at http://newangelesmonthly.com/article.php?id=212&IssueNum=17.

Chapter 6
Santa Monica Sustainable City Plan: Sustainability in Action

Brenden McEneaney

Introduction

The City of Santa Monica is surrounded on three sides by the City of Los Angeles and on the fourth side by the Pacific Ocean. It has 90,000 residents in an area of $21\,km^2$, but the population can swell to 300,000 during the workday, and over 500,000 visitors come to the city in an average weekend. The climate is mild

B. McEneaney (✉)
Office of Sustainability and the Environment, City of Santa Monica, Santa Monica, CA, USA
e-mail: brenden.mceneaney@smgov.net

W.W. Clark II (ed.), *Sustainable Communities*, DOI 10.1007/978-1-4419-0219-1_6,   77
© Springer Science+Business Media, LLC 2010

and temperate, with ocean breezes moderating extreme temperatures year round. Approximately 40 cm of rain falls on the city each year, with almost all of that occurring in the winter months. Effects of climate change and recent extended periods of drought pose a serious threat to the city as well as the region, as most of the potable water is imported from Northern California. The city's location on the oceanfront is as significant for its impact on tourism as it is for its impact on environmental health. Urban runoff is a major cause of pollution in the bay, and during the dry season, 1,890 m^3 of polluted runoff from the Los Angeles Basin enters the city. The city constructed a runoff treatment facility in 2000 to treat the dry season runoff and reuse it for landscape irrigation around town.

The city has a strong liberal political tradition that forms the foundation for many of its progressive environmental and social policies.

Program History

In September 1994, the Santa Monica City Council took steps to address sustainability issues in the community, adopting the Santa Monica Sustainable City Program. The program was initially proposed in 1992 by the City's Task Force on the Environment, a group of community members selected by the City Council as expert advisers on environmental issues. The program was proposed to ensure that Santa Monica could continue to meet its current needs – environmental, economic, and social – without compromising the ability of future generations to do the same. It was designed to help the community begin to think, plan, and act more sustainably – to address the root causes of problems rather than the symptoms of those problems and to provide criteria for evaluating the long-term rather than the short-term impacts of both individual and community-scale decisions.

The 1994 program includes goals and strategies, for the city government and all sectors of the community, to conserve and enhance our local resources, safeguard human health and the environment, maintain a healthy and diverse economy, and improve the livability and quality of life for all community members in Santa Monica. To evaluate the progress toward meeting these goals, numerical indicators were developed and specific targets were set for the city to achieve by the year 2000 in four goal areas:

- Resource conservation
- Transportation
- Pollution prevention and public health protection
- Community and economic development

Progress reports were prepared to track the city's progress toward meeting its goals in 1994, 1996, and 1999. In 2005, a progress report web site was launched, providing up-to-the-moment reporting details.

By 2001, following 7 years of implementation, the Santa Monica Sustainable City Program had achieved much success. Many of the initial targets had been met or exceeded, and Santa Monica had become recognized as a worldwide role model for sustainability. However, in reviewing the progress made since the 1994 adoption of the program, the Task Force on the Environment recognized the need to update and expand the program. They noted that while progress had been made in the right direction, Santa Monica's economy and the activities of its residents, businesses, institutions, and visitors continued to negatively impact human health and the environment. The community was not yet able to provide for the basic needs of all its members. The task force felt that a comprehensive update process to improve and expand the program was necessary to achieve the initial program goals.

Updating the Plan

In reviewing the progress made since the 1994 adoption of the program, the Task Force on the Environment recognized the need to update and expand the Sustainable City goals and indicators to provide a more complete picture of community sustainability and to develop new indicator targets for 2010. The task force felt that a comprehensive update would allow Santa Monica to build on its initial success and to address remaining challenges more effectively.

The update process began in July 2001 with the formation of the Sustainable City Working Group – a large group of community stakeholders that included elected and appointed officials, city staff, and representatives of neighborhood organizations, schools, the business community, and other community groups. The working group met numerous times over the course of 15 months to discuss the myriad issues related to the sustainability of the community. They evaluated the long-term sustainability of Santa Monica using a framework comprised of three forms of community capital that need to be managed with care in order to ensure that the community does not deteriorate. These include natural capital – the natural environment and natural resources of the community; human and social capital – the connectedness among people in the community and the education, skills, and health of the population; and financial and built capital – manufactured goods, buildings, infrastructure, information resources, credit, and debt.

The group proposed significant changes to the initial Sustainable City goals and indicators and assisted with the creation of new indicator targets. Early drafts of the proposed update were revised based on a large amount of public input received during the summer of 2002.

The result of this process is the updated Santa Monica Sustainable City Plan (SCP), which represents the community's vision of Santa Monica as a sustainable city. The change in name from Sustainable City Program to Sustainable City Plan was made to better reflect the long-term comprehensive nature of Santa Monica's vision and the community's efforts to become a sustainable city.

Guiding Principles

The Santa Monica SCP is founded on 10 guiding principles that provide the basis from which effective and sustainable decisions can be made.

1. **The concept of sustainability guides city policy**

   Santa Monica is committed to meeting its existing needs without compromising the ability of future generations to meet their own needs. The long-term impacts of policy choices will be considered to ensure a sustainable legacy.

2. **Protection, preservation, and restoration of the natural environment is a high priority of the city**

   Santa Monica is committed to protecting, preserving, and restoring the natural environment. City decision-making will be guided by a mandate to maximize environmental benefits and reduce or eliminate negative environmental impacts. The city will lead by example and encourage other community stakeholders to make a similar commitment to the environment.

3. **Environmental quality, economic health, and social equity are mutually dependent**

   Sustainability requires that our collective decisions as a city allow our economy and community members to continue to thrive without destroying the natural environment on which we all depend. A healthy environment is integral to the city's long-term economic and societal interests. In achieving a healthy environment, we must ensure that inequitable burdens are not placed on any one geographic or socioeconomic sector of the population and that the benefits of a sustainable community are accessible to all members of the community.

4. **All decisions have implications to the long-term sustainability of Santa Monica**

   The city will ensure that each of its policy decisions and programs is interconnected through the common bond of sustainability as expressed in the guiding principles. The policy and decision-making processes of the city will reflect our sustainability objectives. The city will lead by example and encourage other community stakeholders to use sustainability principles to guide their decisions and actions.

5. **Community awareness, responsibility, participation, and education are key elements of a sustainable community**

   All community members, including individual citizens, community-based groups, businesses, schools, and other institutions, must be aware of their impacts on the environmental, economic, and social health of Santa Monica; must take responsibility for reducing or eliminating those impacts; and must take an active part in community efforts to address sustainability concerns. The city will therefore be a leader in the creation and sponsorship of education opportunities to support community awareness, responsibility, and participation in cooperation with schools, colleges, and other organizations in the community.

6. **Santa Monica recognizes its linkage with the regional, national, and global community**

   Local environmental, economic, and social issues cannot be separated from their broader context. This relationship between local issues and regional, national, and global issues will be recognized and acted upon in the city's programs and policies. The city's programs and policies should therefore be developed as models that can be emulated by other communities. The city will also act as a strong advocate for the development and implementation of model programs and innovative approaches by regional, state, and federal governments that embody the goals of sustainability.

7. **Those sustainability issues most important to the community will be addressed first, and the most cost-effective programs and policies will be selected**
The financial and human resources that are available to the city are limited. The city, with the input of the community, will reevaluate its priorities, programs, and policies annually to ensure that the best possible investments in the future are being made. The evaluation of a program's cost-effectiveness will be based on a complete analysis of the associated costs and benefits, including environmental and social costs and benefits.

8. **The city is committed to procurement decisions which minimize negative environmental and social impacts**
The procurement of products and services by the city and its residents, businesses, and institutions results in environmental, social, and economic impacts both in this country and in other areas of the world. The city will develop and abide by an environmentally and socially responsible procurement policy that emphasizes long-term values and will become a model for other public as well as private organizations. It will advocate for and assist other local agencies, businesses, and residents in adopting sustainable purchasing practices.

9. **Cross-sector partnerships are necessary to achieve sustainable goals**
Threats to the long-term sustainability of Santa Monica are multisector in their causes and require multisector solutions. Partnerships among the city government, businesses, residents, and all community stakeholders are necessary to achieve a sustainable community.

10. **The precautionary principle provides a complementary framework to help guide city decision-makers in the pursuit of sustainability**
The precautionary principle requires a thorough exploration and careful analysis of a wide range of alternatives and a full cost accounting beyond short-term and monetary transaction costs. Based on the best available science, the precautionary principle requires the selection of alternatives that present the least potential threat to human health and the city's natural systems. Where threats of serious or irreversible damage to people or nature exist, lack of full scientific certainty about cause and effect shall not be viewed as sufficient reason for the city to not adopt mitigating measures to prevent the degradation of the environment or protect the health of its citizens. Public participation and an open and transparent decision-making process are critical to finding and selecting alternatives.

Goal Areas

The SCP includes eight goal areas that, taken together, present a vision for sustainability in the community.

Resource conservation
Open space and land use
Environmental and public health housing
Transportation
Community education and civic participation
Economic development
Human dignity

Each goal area is comprised of different target indicators. These indicators not only establish expected levels of performance, but also provide a system on which the program success can be measured. For example, the following are target indicators for the resource conservation goal area:

Resource Conservation

Solid Waste

The city looks at solid waste in terms of the total amount generated, the amount sent to the landfill, and the amount diverted from landfills. The target for generation is to stay at or below the year 2000 baseline through 2010. The target for diversion is to increase the amount diverted to 70% of the total generated by 2010.

Water Use

The target for consumption is to reduce the overall water use by 20% of 2000 levels by 2010. In 2000, water use was 13.4 million gallons per day (MGD); a 20% reduction in that usage is 10.7 MGD. The city aims to increase the amount of locally obtained potable water to its 1995 high point of 70% of total water use by the year 2010. The city is also working to maximize nonpotable water use when appropriate.

Energy Use

This indicator tracks historical energy use, both as a benchmark for energy conservation and for improvement of other measures: renewable energy and greenhouse gas emissions. No target has been set for energy consumption, though obviously energy use reduction is desirable. The residential, commercial, industrial, transportation, and solid waste sectors make up citywide energy use.

Renewable Energy

The target for renewable energy use in the city is 25% of citywide electricity use (grid electricity) from renewable sources by the year 2010. In addition,

1% of all electricity used should come from clean distributed generation by 2010.

Greenhouse Gas Emissions

The target for this indicator is to reduce emissions 30% below 1990 levels by 2015 for city operations and 15% below 1990 levels by 2015 for the city as a whole.

Ecological Footprint

The city does not have a target set for the ecological footprint, but a downward trend in the size of our footprint is desirable.

Sustainable Procurement

The target for this indicator is that by 2010, 20 purchased product categories will be converted from conventional to environmentally preferred products. Each year the staff will target five product categories per year converted.

In 2007 the targets were as follows:

Food packaging
Carpet
Computers
Copy and printing paper
Janitorial paper supplies (toilet tissue, paper towels)
Office supplies

Green Construction

The goal is to certify all new buildings eligible for LEED certification that are 10,000 square feet or larger, by 2010. Of these, 20% should attain LEED Silver, 10% Gold, and 2% Platinum certification, with the remainder categorized simply as certified. Further, 50% of new buildings smaller than 10,000 square feet shall obtain at least LEED certification or its equivalent by 2010. This target includes all municipal construction.

Governance

The City's Task Force on the Environment provided leadership on behalf of the community for the SCP since the plan was adopted in 1994. With the update and expansion of the SCP into new and more diverse goal areas, the Task Force on the Environment recommended the creation of a Sustainable City Task Force (SCTF).

The SCTF includes broad representation from community stakeholders with expertise in all of the SCP goal areas to guide the program in the future. The City Council created the SCTF and appointed 11 members of the community with expertise in the various areas of sustainability on March 23, 2004.

An interdepartmental advisory team, chaired by a representative from the city manager's office, was created to coordinate internal city activities so they are consistent with the SCP goals and to help facilitate the future implementation of innovative programs and policies. Members of this group serve as SCP liaisons to their respective departments.

The SCTF and the Sustainable City Advisory Team are both responsible for developing a comprehensive implementation plan for meeting Sustainable City goals and targets and for coordinating implementation, both interdepartmentally and within community stakeholder groups.

Report Card

The Sustainable City Report Card was developed in order to give the community a quick-glance summary of the city's performance. The report card measures the entire community's progress toward meeting the SCP goals, which were developed by the community during a 15-month process in 2001–2002 and adopted by the City Council in 2003.

The report card grades the performance of the entire community. The grades were developed based on a detailed analysis of indicator data found in the Sustainable City Progress Report. The city employees draft grades initially with the assistance of an outside consultant and outside experts, including representatives of the City's Task Force on the Environment and SCTF. The final grades presented in the report card receive outside review and approval by experts in various public processes.

The grades in the report card have two aspects. First, they grade the city's performance toward our indicator targets. A second grade was added to evaluate the level of effort toward those targets. The addition of the second grade was intended to reflect the impact of outside influences on the indicator targets. For example, the health of the Santa Monica Bay depends largely on runoff that arrives from all over the Los Angeles Basin. Although the city has implemented many mitigation measures, much of the impact is beyond our control. The second grade was also established as a way to recognize the city's efforts, aside from the performance toward the target indicators.

The process of evaluating both the efforts and the performance of the city toward its sustainability goals is an essential component. Rather than presenting a rosy, politically acceptable picture, the report card is an honest evaluation of the state of the local environment. The mediocrity of some of the grades gives the city more incentive to develop innovative programs to address these issues. While the effort grades are a way to recognize the city's efforts in the face of many mitigating factors, the goals must ultimately be achieved for the city to achieve its vision of

SUSTAINABLE CITY REPORT CARD

RESOURCE CONSERVATION		2005	2006	2007	2008
	Grade	C	C+	C	C+
	Effort	A	A	A-	A-

Goals: Decrease consumption of non-local, non-renewable, non-recyclable energy, water, materials and fuels / Promote renewable resource use

Santa Monica is a leader in programs and policies related to resource conservation. Residents and businesses prioritize resource use reduction, but greater effort is needed to meet Sustainable City Plan targets. Solid waste generation exceeds the Sustainable City Plan ceiling and continues to increase. Solid waste diversion improved this year to 68% and is approaching the 70% target. Total water use is down by 6% and the percent of locally obtained water improved. This trend is expected to continue as MTBE remediation work begins. Despite aggressive energy conservation measures, electricity and natural gas use have increased. Currently 18% of citywide energy is from renewable sources. Solar Santa Monica continues to deploy energy efficiency, solar power and clean distributed generation in the community. There are 139 grid connected solar projects in the city representing 926 kilowatts of solar capacity. The city continues to purchase 100% renewable power for municipal operations and the commitment to green building resulted in the Civic Center Parking Structure receiving Leadership in Energy and Environmental Design (LEED) certification. Despite the increases in solid waste generation and energy use, the grade improvement reflects successes in municipal sustainable procurement, citywide solid waste diversion, water use reduction, and green building.

ENVIRONMENTAL & PUBLIC HEALTH		2005	2006	2007	2008
	Grade	B	B-	C	C
	Effort	A	A	A-	A-

Goals: Minimize or eliminate the use of hazardous and toxic materials and the levels of pollutants entering the air, soil and water

Last year Santa Monica voters passed the Clean Beaches and Ocean parcel tax to fund the Watershed Management Plan. Implementation of this comprehensive 20-year approach to improving water quality in the Santa Monica Bay began this year. Despite efforts to eliminate flow to the beach, there was a 31% increase in the number of posted beach warnings during the dry season. Research is underway to identify the cause of beach postings. In an effort to reduce marine debris, the city implemented a ban on the use of all non-recyclable plastic take-out containers. The Council also authorized staff to develop a ban on plastic bags. The amount of urban runoff captured and treated at the Santa Monica Urban Runoff and Recycling Facility increased 9% to 33 million gallons. The cumulative number of households properly disposing of hazardous waste at the Household Hazardous Waste Center increased from 36% to 43 percent. The amount of e-waste collected citywide doubled from 21 tons last year to 42 tons this year. Four thriving farmers' markets, one of which is regularly a zero-waste event, provide access to fresh, locally grown and organic produce. Despite these successes, the grade remains consistent because the city is far from reaching its targets for Santa Monica Bay health.

Fig. 6.1 Report card

sustainability. No matter how admirable the effort, the end result must be fewer kilowatt-hours produced by fossil fuels, better water quality in the Santa Monica Bay, and fewer greenhouse gas emissions.

To this end, through the revision and reevaluation process, it was observed that some of the city's sustainability goals were not consistent with the real reductions in energy use, material use, water use, and waste generation. An example is the target for green construction. The city set a target of 100% of newly constructed buildings achieving LEED certification from the United States Green Building Council by the year 2010. As of late 2008, fewer than 2,000 total buildings in the history of the USGBC had been LEED certified. Beyond the incredibly ambitious performance level set by this target, the LEED certification is a proxy for other more meaningful indicators, like energy efficiency and water use reduction. Indeed, a building can be LEED certified without achieving any water use reductions at all[1].

---

[1] As of the writing of this chapter, the LEED rating system is undergoing a revision process, and one of the proposed changes is to require mandatory water use reduction in all certified buildings.

Santa Monica Green Building Program

While the SCP sets the policy roadmap for how Santa Monica journeys toward being a sustainable city, a wide variety of programs are implemented in the city to achieve those policy goals. From affordable housing to transportation management, these programs use the SCP to frame their short- and long-term decisions. One program that this chapter will focus on is the Green Building Program. This program is particularly interesting because it addresses a wide range of sustainability goals simultaneously. While it can be argued that the primary goal of green building is resource conservation, green building strategies work to improve public health, employee productivity, student performance, and access to public transportation and open space and generally help make healthy, vibrant communities.

History

In 1999, Santa Monica published *Commercial Green Building Guidelines*. This document compiled a list of suggested measures, organized in several categories, to help a builder improve the environmental performance of a new building's design. This guideline was one of the first of its kind, drawing on knowledge from city employees and from experts in the field. That year, the city also required energy efficiency improvements for commercial construction beyond what was required in the California Building Code.

To manage the establishment of green building code requirements and to improve outreach and education about green building practices, the city created the position of Green Building Program Advisor in 2002. With a dedicated staff member to oversee the program, the city instituted grants for LEED-certified buildings, grants for innovative technologies addressing urban runoff or energy efficiency, and expedited plan review for buildings pursuing LEED certification.

The city followed up its first commercial green building guidelines with the publication of the residential guidelines in 2005. Subsequently, Santa Monica has also expedited permitting and waived permit fees for solar installations.

Program

The City of Santa Monica Green Building Program involves four elements: education, motivation, facilitation, and regulation.

Education

Perhaps the first and most important role that a municipality can play in furthering goals of sustainability is to provide education to the public. Using principles of sustainability as a lens through which to view policy decisions is a new concept for many cities, and by educating the community, the City of Santa Monica has provided a transparent decision-making process. Similarly, as the Green Building Program puts the SCP into action, education about green building concerns is central to the program's mission.

The city does outreach through many avenues to educate the community about green building. It has partnered with a local nonprofit organization to establish a Green Building Resource Center. This is a storefront space in the center of town that is stocked with examples of green materials and publications for research. The staff there can also give advice and consultations on building projects that visitors may have. Each year, the city also hosts a green building conference and expo that is free and open to the public. At this event, known as AltBuild, vendors showcase their materials and services, and the focus is on both local suppliers and innovative products. Attendees can also hear from experts in the field on a variety of topics related to green building.

The Green Building Program web site[2] is one of the most highly trafficked web sites in the city. The guidelines for commercial and residential construction that had previously been developed have been put online in electronic format. The web site also contains a design advisor that creates a list of suggested measures based on information about a project input by the user. As of late 2009, the design advisor was removed from the website in favor of more general guidelines for green construction.

Motivation

The city has established several incentives to encourage exemplary performance in green construction. All new buildings applying for LEED certification can receive expedited plan review, which can reduce the amount of time required to get a building permit. Applications to install solar panels receive the same expedited plan review, and in addition, no permit or inspection fees are charged. Projects that are LEED certified and that meet other occupancy criteria are allowed certain development incentives, such as density and FAR increases in allowable density.

In addition, the city has made Green Building Grants available to buildings that achieve LEED certification. The following figure shows the details of the grant program.

The city also created a grant program to offer incentives for innovative technologies that address either urban runoff pollution or energy efficiency. The city will cover up to 50% of the cost of the technology, that is, up to $5,000. A parallel grant program exists for landscape water use efficiency up to an amount of $20,000. Though the grants are not administered through the Green Building Program, the reduced landscape water use certainly supports the resource conservation goals in the SCP. As of August 2009, all of the available grants have been awarded. As LEED Certification is now more widely attainable (Santa Monica has one of the highest numbers of LEED Buildings per capita), the grant program has been closed.

---

[2] www.smgreen.org.

Green Building Grant Amounts			
	Single Family	Multifamily (up to 10 units)	Commercial
Certified	$3,000	$2,000/unit	$20,000
Silver	$4,000	$2,500/unit	$25,000
Gold	$6,000	$3,000/unit	$30,000
Platinum	$8,000	$3,500/unit	$35,000

Facilitation

This element of the Green Building Program is indirect and often difficult to measure; however, it remains an important element. The creation of a permanent position for a Green Building Program Advisor also created a permanent advocate for green building strategies and technologies. As such, the Green Building Program Advisor can work behind the scenes to remove impediments to those strategies and technologies. Existing building and zoning codes can present legal and administrative obstacles for pursuing strategies for energy and water use efficiency. The advisor can develop code revisions as well as suggest improvements for administrative processes that would help encourage best green building practices.

Regulation

Beyond the incentives to encourage exemplary performance, the City of Santa Monica has established a minimum level of environmental performance that is expected of building projects in the city. Codifying this performance in the building code creates not only a clear expectation for development but also a level playing field for any cost hurdles that builders might face. Furthermore, it ensures that all building occupants would enjoy and benefit from cleaner, healthier, more efficient buildings, not just those wealthy enough to afford a high-performance building. This has been one criticism of the green building trend in the United States, particularly in the residential sector – that green is expensive. Recent studies are showing that there is no price premium for building green, but prices may be affected by the perception of value attributed to green buildings. Under the SCP, environmental, economic, and equity concerns are all important, and regulatory mandates are an essential tool for cities to address these concerns.

All new building projects are required to comply with the Green Building Ordinance.[3] This ordinance addresses energy efficiency, green construction materials, landscape water conservation, and construction and demolition waste management.

---

[3] Santa Monica Municipal Code Chapter 8.108.

Energy Efficiency

All new buildings in the city must use 10% less energy than the State Energy Code requires. The California State Energy Code is already one of the most stringent in the country, and the City of Santa Monica requires better performance that can be shown either with an energy model or with a set of prescriptive requirements.

Green Building Materials

To promote the use of environmentally friendly construction materials, all new buildings are required to use a green option in five areas of construction. These five areas are chosen from a list of 10 areas developed by the Green Building Program. In each area, several common green options are listed, and project teams have the option to suggest an alternative material that is not on the list. The list includes things like zero-VOC[4] paint or FSC[5]-certified framing lumber or concrete with a high fly ash content.

Landscape Water Conservation

Southern California is a region with low rainfall. In addition, the potable water in the region comes from areas far to the north. Pumping water over this long distance requires enormous amounts of energy, and climate change has significantly reduced the amount of rainfall and snowpack, threatening long-term supplies to the region.

Much of the water use associated with buildings happens outside in the landscaping (approximately 50% of the use in single-family residential buildings). New buildings may plant only 20% of the landscaped area of the site with high-water using plants. This includes traditional turfgrasses, or sod. Plants that are invasive to Southern California are not permitted. Irrigation systems must have high efficiency with a precipitation rate of less than 0.75 inches per hour. Drip irrigation must be used for plants that are larger than 1 gallon in size, and no spray irrigation can be located within 18 inches of hardscapes to prevent the urban runoff pollution that inevitably comes from such sprinklers.

Construction and Demolition Waste Diversion

The State of California requires that municipalities achieve a 50% diversion rate for all construction and demolition waste. While not all cities have reached compliance with this requirement, Santa Monica requires a 65% diversion rate. In the calculation of this rate, inert materials like soil cannot be included, as their large quantities tend to distort the actual amount of waste diversion.

---

[4]Volatile organic compound.

[5]Forest Stewardship Council: a nonprofit organization that certifies forests and forest products as being sustainably managed and harvested (www.fsc.org).

These requirements apply to all new buildings in the city. Furthermore, the city has led by example, by requiring that all municipal construction projects achieve a LEED Silver certification or higher. This requirement not only benefits taxpayers by reducing the operating costs of city facilities over their lifetime, but also serves as another educational tool for demonstrating green building strategies.

Fig. 6.2 Main library
Photo credit: www.amywilliamsphoto.com

Solar Santa Monica

History

In the Fall of 2006, the Santa Monica City Council approved a plan for a program called the Community Energy Independence Initiative (CEII). The city had previously gotten a US Department of Energy Grant to do a citywide study on the potential for rooftop solar generation. This study used satellite imagery to look at roofs of buildings in the city. The roof size, location, orientation, shading, and other factors were taken into account. The study found that when combined with sweeping energy efficiency measures, the city could actually produce all of its energy needs with solar photovoltaic and solar water heating on rooftops in the city. This result led to the development of the CEII, now known as Solar Santa Monica.

Fig. 6.3 Parking structure

Program

Solar Santa Monica is a city-run program that aims to get solar on every available rooftop in the city. Program participants first receive a free assessment of their home or business. This assessment gives the participant recommendations in key areas where energy efficiency can be improved. This is a critical element of the program. The city has a vested interest in reducing total energy use, not just maximizing the amount of solar that can be installed. This effort toward energy efficiency saves residents money on their utility bills, and the investment in efficiency is typically smaller and has a much higher return than the investment in solar panels.

During the site assessment, the roof of the building is also evaluated for its solar potential based on the same factors that went into the study – shading, orientation, roof pitch, and so on. When the participant gives information about their utility bills, the Solar Santa Monica assessor can give a report that summarizes the energy efficiency measures – what they will cost and how much energy can be saved – and an estimate on what it would cost to install solar power.

This report is given to residents regardless of how or when they choose to install solar. Solar Santa Monica supports residents, no matter which contractor they choose, but an additional benefit provided by the program is a partnership with several local installers. The city has reviewed applications from qualified installers and has established partnerships with several of them. The Solar Santa Monica assessor can refer a resident to one of these prequalified installers. This way, a resident who might be unfamiliar with any solar installers, or unfamiliar with their qualifications, can be put in touch with an installer who has previously been vetted. In addition, much of the background work in evaluating the solar potential of the site has already been done, reducing costs for the installers.

Results

Since the program inception, the city has approximately doubled the total installed solar capacity. Much of this increase in capacity has come from a couple of large installations, but the number of smaller residential systems has also increased dramatically.

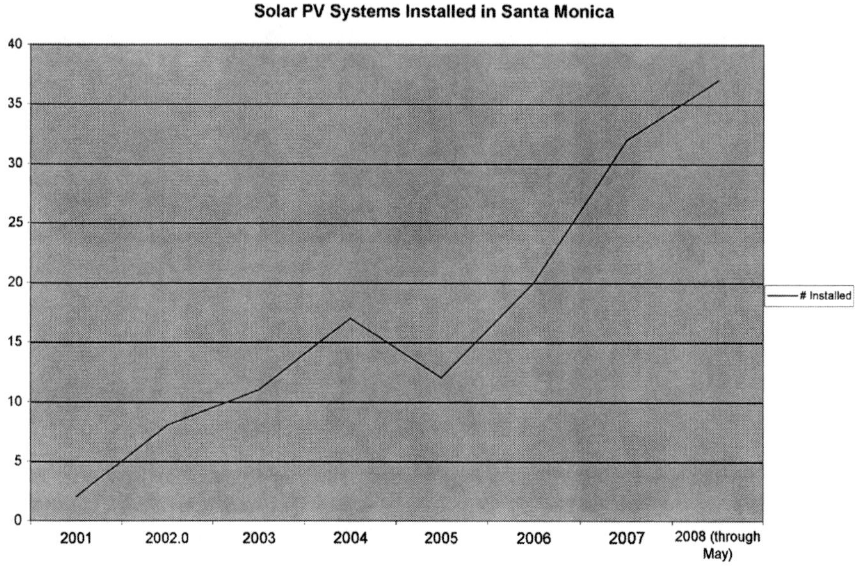

Fig. 6.4 Graph of number of installations

To date, most of the installations have not come through the solar installers that the city partners with. This was a surprising result, but it seems that the efforts of the program to get out in the community have generally raised awareness about solar and energy efficiency, and so Solar Santa Monica has had a compound effect much greater than any direct incentive, such as a rebate on solar installations. The Solar Santa Monica team has

- staffed a booth at farmers' markets around town;
- made presentations to dozens of homeowners' associations and neighborhood organizations;
- organized several Solar Town Halls to communicate important information about tax credits and legislative changes;
- given out information and the city's list of installers to non–Santa Monica residents;
- left pamphlets on the doorknobs of 6,000 homes in the city;
- identified the "100 best roofs" for solar installation and contacted the building owners;
- made presentations to the local chamber of commerce.

The city has about 1 MW of installed capacity using rooftop solar. This represents approximately 1% of the city's electricity demand. While this is a modest result, the recent increase in the number of applications to install solar energy systems is promising. Furthermore, the federal tax credits available for solar energy have been extended and increased, and so more growth in solar installations is anticipated for 2009.

Conclusion

The City of Santa Monica has a progressive plan to steer itself onto a sustainable path. The SCP is a policy vision with ample community and stakeholder input. This plan guides decision-making in the city. The plan was not established as an immutable dogma, but rather as a living document that sets out expectations, requires measurement of performance, and revises the vision of sustainability accordingly. Having a plan like this in place allows the city to establish broad-based programs like the Green Building Program as well as targeted efforts like the Solar Santa Monica program. Having a clear vision with support from the community helps the success of these programs as individual strategies to move toward a more sustainable city.

Chapter 7
Renewable Energy Practices in the City and County of San Francisco

Jatan Dugar, Phil Ting, and Johanna Partin

San Francisco has always been a national leader when it comes to developing environmental-friendly practices and lifestyles. With a population of about 800,000 people, San Francisco is the fourth largest city in California. The city is known as the political, cultural, and financial center of the Bay Area, which is home to more

J. Dugar (✉)
University of California, Los Angeles, USA
e-mail: jatandugar@googlemail.com

Jatan Dugar. With contributing sources: Phil Ting and Johanna Partin.
Phil Ting, Assessor-Recorder for the City of San Francisco, is the contributing author, and Johanna Partin, Renewable Energy Program Manager for the San Francisco Department of the Environment, is the contributor.

W.W. Clark II (ed.), *Sustainable Communities*, DOI 10.1007/978-1-4419-0219-1_7,   95
© Springer Science+Business Media, LLC 2010

than 7 million people. Historically, San Francisco has attracted forward-thinking minds from diverse cultural backgrounds that helped to mold it into the progressive, accepting community that it is today.

The city and county of San Francisco is situated on a peninsula, surround by coasts on three sides. The city is well known for its moderate temperature: A Mediterranean climate that remains around 57°F (13.8°C) year round. The enviable weather, however, comes at a price. The geography of San Francisco is susceptible to the effects of climate change on a large scale. Threats include a rise in sea level and a reduction in the amount of snow pack for the water supply and hydroelectric power. The city's water supply comes from the Sierras, which contributes to the bulk of the 190 MW of renewable energy that the city uses for power.

In this instance, renewable energy includes any source that is not carbon based such as solar power, hydropower, wind power, biofuels, and geothermal energy. This chapter will include a brief summary of each energy source and the present and future renewable energy objectives of the city of San Francisco.

Solar Power[1]

San Francisco is a national leader in the development of renewable energy on city-owned buildings. However, it is falling behind in the development of privately owned sources of renewable energy on local homes and businesses.

Installing solar (photovoltaic) energy systems in San Francisco costs almost 10% more than the Bay Area's average cost of solar energy: almost $10/W compared to the $9.36/W Bay Area average. This cost differential is part of the reason why San Francisco ranks last among the 10 Bay Area counties in terms of solar kilowatts installed per capita. In order to meet the goal of making solar energy more affordable for all San Franciscans who wish to install systems on their homes and businesses, the core economics need to be addressed – the average installation (factoring in commercial and residential averages) costs $30,000 in San Francisco. Although state rebates (administered by PG&E) and federal tax incentives can reduce these costs by 30%, the city believes it could do more. Also considering San Francisco enjoys almost double the average hours of sun per day as Berlin, which is recognized as a world leader in urban solar generation, doing more is well within our grasp.

The San Francisco Solar Task Force, cochaired by Assessor-Recorder Phil Ting and Vote Solar Founder David Hochschild, has met since February 2007 with the goal of identifying an action plan to increase solar installations in the city. The task force concluded that a direct cash incentive, paired with a low-interest loan program, could utilize a small amount of city funding to deliver an unprecedented amount of privately owned solar energy to our city.

---

[1] Ting, Phil. "Go Solar SF." *San Francisco's Groundbreaking Solar Incentive Program*. November 24, 2008.

When the task force first convened, there were some 580 solar installations in San Francisco and just over 5 MW of solar power. As of February 2009 (2 years later) there are 1,040 solar installations and a total of 7.2 MW of solar power in the city.

Incentive Program

San Francisco's solar energy incentive program, officially called "GoSolarSF," was launched on July 1, 2008. It is among the largest municipal solar incentive programs in the country. Mayor Gavin Newsom, Assessor-Recorder Phil Ting, and the San Francisco Solar Task Force all supported this legislation. The task force, assembled in February 2007, is a coalition of individuals representing labor, environmental advocacy, solar industry, business, and local and state governments. The 10-year program consists of one-time incentive payments made on an upfront payment to residential, commercial, and nonprofit property owners who install rooftop solar energy systems in San Francisco. This program involves a simple and direct payment to local homes and businesses, and a multi-year funding commitment to support these solar incentive payments.

GoSolarSF is a city-backed financial incentive program that will provide between $3 and 6,000 to homeowners and up to $10,000 to business owners that install solar panels on their properties. The incentive levels will reduce total costs of the average residential installation by 50%. The incentives are mutually exclusive and structured as follows:

- $3,000 is the baseline level of assistance for residential installations
- $4,000 for solar installations completed by installers with offices in San Francisco, in order to help build a sustainable local solar industry and deliver "green jobs" to local residents.
- $5,000 incentive for installations completed in Environmental Justice Districts (communities most adversely affected by the city's power generation) and other residents that are at or below the Area Median Income
- $6,000 incentive for installations completed using workers from the city's workforce training programs
- $1500/kW installed on commercial properties (maximum payment of $10,000)

GoSolarSF is expected to generate 50 MW of new solar energy over the next 10 years, thereby helping the city attain its established goal of reducing greenhouse gas emissions to 20% below 1990 levels by 2012. San Francisco will also be doing its part in meeting the solar goals that the state set in 2006 with the creation of CSI that include the installation of 3,000 MW of solar electric power on the equivalent of a million roofs in California by 2017.

There is also a complementary low-income solar pilot program. The 1-year pilot, crafted by Supervisor Ross Mirkarimi, allocates another $1.5 million to buildings owned and operated by nonprofit organizations and low-income single and multifamily residential applicants.

A San Francisco solar customer that receives the city's solar incentive would be eligible for both the state's incentive program and the federal solar tax credit. In order to keep the city's solar incentive program application process simple, solar property owners who qualify for the state rebate program also qualify for the city's incentive. However, the proposed ordinance also directs the program administrator to establish simple eligibility criteria for all applicants, even those for whom the state's incentive program is not available.

More information on the program and the number of applications to date can be found at www.sfwater.org/gosolarsf.

Expedited Permit

Although 60% of all solar installations in San Francisco were able to take advantage of the expedited permit process that the city had in place, the remainder had to go through a process that could take months to complete and ended up being more costly.

This difficult process added to the costs of residential solar in San Francisco – it was costing on average $2,500 more to install solar in San Francisco than in San Rafael and $1,000 more than in Oakland – primarily due to the complexities associated with our permitting process.

The task force understood that more could be done to make solar permitting easier for both the applicant and permitting authorities, while maintaining an appropriate level of oversight in solar installations. After a few months of meeting, the Department of Building Inspections revised the standard permitting process, resulting in most electrical PV applications being issued in a matter of weeks rather than months.

At $170 (they were $111 and were raised in the last budget cycle), San Francisco has among the lowest permit rates in the state; a Sierra Club study reported last year that the average permit fee is $282 to install a 3 kW rooftop solar system in a single-family home.

Clean Tech Payroll Tax Exemption

Another policy initiative that the task force pushed for was the expansion of the San Francisco's Clean Technology Payroll Tax Exclusion to include solar installers. San Francisco continues to be at a disadvantage in the solar energy market compared to the neighboring counties because the cost of doing business is higher, resulting in higher costs to install solar and many fewer installations per capita. Allowing solar installers to receive the payroll tax exemption would help even the playing field and allow for additional green collar jobs in San Francisco. The Board of Supervisors approved legislation enacting this expansion of the tax credit in December 2007.

Solar Monitoring

The San Francisco Public Utilities Commission installed a solar monitoring system to track the amount of solar power available in different locations throughout the city. In addition, the Department of the Environment created an up-to-date solar map of all current projects in the city, known as SF Solar Map. The web site tracks all municipal, commercial, nonprofit, and residential solar installations as well as solar monitoring stations. The site can also tell users the amount of solar PV potential and average electricity savings of any specific address that is entered into the system. It was created by the Department of the Environment to help people understand whether their property has solar potential. The general consensus among San Franciscans seems to be the disbelief in solar potential for the city – perhaps due to the ubiquitous fog. By utilizing web sites such as sf.solarmap.org, this mentality will hopefully be shifted and more people will be encouraged to install solar systems on their residences or businesses.

Fig. 7.1 San Francisco Solar Map http://sf.solarmap.org

Mayor's Solar Founders Circle

The office of San Francisco Mayor Gavin Newsom and the Department of the Environment created the Solar Founders Circle in September 2008 in order to encourage new businesses to go solar. The program offers free solar assessments and energy efficiency audits if they sign up to be a part of the Founders Circle. The mayor has already established a goal for businesses to install 5 MW of solar power by September 2009, doubling the current amount of solar power generation. San Francisco businesses and nonprofits have the capacity to install more than 170 MW of solar panels on their roofs.

The Moscone Convention Center Solar Project

The Moscone Center in downtown San Francisco is the largest exhibition complex in the city. In 2004, The PowerLight Corporation installed solar panels on the roof of the center, creating 670 KW of electricity for the building. The San Francisco Mayor's Energy Conservation Account (MECA) directed funding for the solar program, and the program also received rebates from the California Public Utilities Commission and California Energy Commission.

Fig. 7.2 Solar panels on the roof of the Moscone Center

Wave Power

San Francisco's prime geography allows for the possibility of offshore wave power to be used as a renewable energy source. Offshore wave power takes advantage of the circular motion of ocean waves. Underwater devices collect the circular energy of a wave, which is then transported to the energy grid. The utilization of wave power is imperative in order to reach San Francisco's goal of using 100% renewable energy, and it is expected that the majority of renewable energy will come from ocean power in the near future.

Originally, the city of San Francisco was interested in studying tidal or current power. Tidal power takes advantage of the back and forth motion of ocean currents underneath the water. Energy is harbored by small turbines under the water, similar to wind turbines. Tidal power is usually advantageous because you only need small systems to extract power. The ocean water underneath the San Francisco Golden Gate Bridge is one of the largest, fastest flowing currents in the Bay Area. Because of the advantageous conditions, the Electric Power Research Institute (EPRI) conducted studies in 1995 to find out the average extractable power of the currents. Average extractable power refers to how much power you can take out of the tides without having an adverse impact on the environment. The study conservatively estimated that the Golden Gate site has 35.5 MW of total extractable average annual power and that on average 15–17 MW of this power could realistically be extracted by technologies currently in development. This average, however, was recently rebuked by a study in 2007, which confirmed that the area only had average extractable power of 1.5 MW. Therefore, efforts to pursue current or tidal power have been halted in favor of wave power.

After researchers at the Department of the Environment realized that wave power would be more advantageous in terms of the amount of energy produced, they decided to focus their attention solely on wave power. The department began a feasibility study in 2008, which proved to be very promising only 6–8 months after its initiation. The studies were so effective that the department decided to submit a preliminary permit to the Federal Regulation and Oversight of Energy Committee (FERC). If the permit is accepted, the city of San Francisco will be granted the rights to conduct the necessary studies to implement a pilot project in the San Francisco bay. Most of the studies that are required to take place concern the effects of a wave power system on the surrounding environment. Underwater systems have the potential to cause interference to aquatic life, as well as to fishing and shipping routes. If all of the studies prove that installing a wave power system would not, in fact, harm the environment, then the city of San Francisco will be allowed to submit a license to actually install an underwater system. The entire process is estimated to take about 3 years to complete.

There are a variety of ocean power systems available today. For example, surface energy systems such as buoys or absorbers create mechanical energy which is converted to electricity and then sent to a power line. Another technology is the Wave Dragon, which creates a walled-in area in the ocean to focus waves onto a ramp,

which is located on an offshore reservoir.[2] For example, the CETO system consists of a piston pump with an attached float anchored to the sea floor. The waves
cause the floats to rise and fall, creating pressurized water which is transported to an
onshore generator used for hydraulic power and reverse osmosis desalination processes.[3] All of these systems will be examined carefully by the Department of the
Environment over the next 3 years, and selection will be determined based on which
system works best with the surrounding ecosystem.

Although the wave power studies are not yet conclusive, the Department of the
Environment estimates anywhere from 30 to 100 MW of average energy extracted
from wave power systems. The amount of energy harvested from ocean power may
actually surpass that of solar power in the near future.

CCSF Geothermal Project

The City College of San Francisco created a project in 2007 to install a 700-ton
Geothermal Heat Exchange System at Ocean Campus. The project, scheduled to be
finished in 2010, is one of the largest geothermal projects in Northern California.
The site in the Balboa Reservoir of Ocean campus will be used in conjunction with
San Francisco State University students. The multipurpose building will be used for
classrooms, offices, visual and performance arts centers, and a technology center.

The construction is managed by Bovis Lend Lease Inc., Timmons Design
Engineers, VBN Architects, EnLINK, and Proven Management and has an estimated construction budget of $7,300,000. The project has caused a lot of excitement
and anticipation because of the advanced geothermal heating and cooling systems
that will power the building. Geothermal systems use underground air to cool or
heat water that is pumped downward and then back up above the ground. The
earth's surface maintains a constant temperature between 50 and 60°F (10 and 16°C)
regardless of the outside temperature. During cooler temperatures, cool air is sent
underground to the relatively warmer earth and is transferred into hot air by means of
a heat pump system. The system uses underground air to cool water that is pumped
above and beneath the ground. The water pipes are connected to a central heating
and cooling system for the building.

This process is considered to be renewable energy because it does not require
any fuel to function. The constant temperature of the earth allows for an efficient,
reliable heating and cooling system that does not emit any by-products. According
to Bovis, the CCSF building will be certified with a LEED rating of silver.[4] The
construction of the new CCSF building is currently being used as a model for other
business and residences to use efficient, renewable sources of heating and cooling
on their buildings.

---

[2]http://www.wavedragon.net/index.php?option=com_frontpage&Itemid=1.

[3]http://www.carnegiecorp.com.au/index.php?url=/ceto/ceto-overview.

[4]http://www.sfbuildingtradescouncil.org/index2.php?option=com_content&do_pdf=1&id=287.

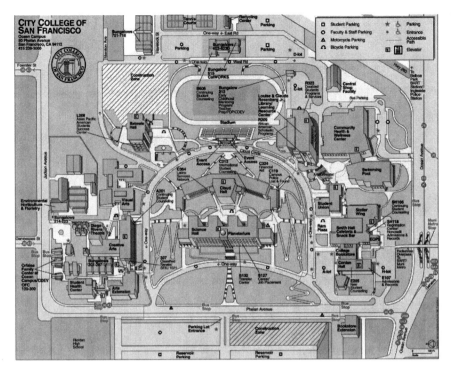

Fig. 7.3 City College of San Francisco, Ocean Campus Map (Credit: SF City College)

Fig. 7.4 Construction of geothermal ground loop system to provide heating and cooling (Credit: Bovis Land Lease)

Fig. 7.5 Joint Use Facility rendering (Credit: VBN Architects)

Wind Power

In April 2008, San Francisco Mayor Gavin Newsom announced the formation of a task force called the Urban Windpower Task Force dedicated to the establishment of wind power in the city. Although there are currently no wind turbines on San Francisco municipal buildings, Newsom is committed to a short-term goal of 50 MW of renewable energy by 2012, and the creation of wind turbines will certainly add to this effect.

Hunter's Point, also known as Bayview in southeastern San Francisco, is a possible site for wind turbine construction, as its seaside location proves to be favorable for wind speeds. Discussion has also been made regarding the possibility of Candlestick Park. In addition, the Windpower Task Force is considering the option of offshore wind farms, to produce a maximum of 420 MW of electricity[5] from 130 open-water wind turbines.

The offshore wind potential on the western side of the city and county of San Francisco is indicated to be Wind Power Class 3 to Wind Power Class 4. This is equivalent to a 50-m wind speed of 6.4–7.5 mps. To confirm this assessment, the wind speed and wind direction data from the stationary buoy off the coast of San Francisco is analyzed.

One of the sites in the CEC Study of the wind resource in the city and county of San Francisco was located at Hunter's Point. The results of that study indicated

---

[5] http://www.examiner.com/a-1514240~The_magic_of_wind_power.html.

a 10-m annual average wind speed of 8.4 mph, nearly 20.5% lower than the annual average wind speed recorded at San Francisco International Airport. The expected wind resource for most of this area would be Class 1; however, a Class 2 or Class 3 resource may exist on the hill west of the stadium complex at Bayview Park and the communication tower location.

These projects have been hindered by slow movement at the federal level as well as a lack of sufficient funding for small-scale projects including residential facilities. As Johanna Partin stated in the *San Francisco Examiner*, "the city is considering subsidizing the prices of so-called wind anemometers, which retail for $150, or renting them out to residents to help defray the costs of measurement devices."

Currently, there are six functioning wind turbines in San Francisco: one 45-foot vertical-axis windmill on a residence in the Mission, a pair of horizontally rotating turbines on a garage in the Castro, a 7-foot turbine on a Bernal Heights roof, a vertical axis turbine at the Randall Museum, and a pilot turbine on Treasure Island.

Recycling and Composting

San Francisco is well known for its extensive recycling and compost programs that are widespread throughout the city. The fact that the city "diverts 69 percent of its waste and discarded materials from landfill"[6] is primarily a result of the population's collective awareness to "do the right thing." In 2004, the city adopted the Zero Waste initiative, designed to reduce waste by maximizing recycling and reusing practices. The Zero Waste California program "requires that we redefine the concept of 'waste' in our society. In the past, waste was considered a natural by-product of our culture. Now, it is time to recognize that proper resource management, not waste management, is at the heart of reducing waste sent to landfills."[7] The Zero Waste resolution of 2004 had a goal of 75% waste diversion by 2010 and zero waste by 2020. The goals are based on the current incentives such as "pay as you throw" plans, where your garbage bill is lowered according to how much you recycle.

In addition to promoting residential recycling, the city of San Francisco also endorses the use of sorted recycling bins at all events that are held in a city park. San Francisco Special Events Ordinance No. 73–89 "requires any applicant seeking permission for the temporary use or occupancy of a public street, a street fair or an athletic event within the city and county that includes the dispensing of beverages or which generates large amounts of other materials to submit a recycling plan."[8]

---

[6]http://www.sfenvironment.org/our_programs/interests.html?ssi=3&ti=5&ii=14.

[7]http://www.zerowaste.ca.gov/WhatIs.htm.

[8]http://www.sfenvironment.org/our_programs/interests.html?ssi=3&ti=5&ii=191.

San Francisco is also known for its composting program, which was one of the first programs of its kind in the United States. San Francisco residents and business contribute more than 300 tons of compost per day,[9] which is sent to the Jepson-Prairie compost facility. The compost creates a fertile soil used for agricultural purposes throughout California.

Other Practices

Hydropower

Eighty-five percent of San Francisco's total water needs come from the Hetch Hetchy reservoir, in Yosemite Valley, which is used to power all city facilities such as city hall, airports, hospitals, and schools as well as water for 2.3 million people. The Raker Act of 1913 requires the city to sell power generated from Hetch Hetchy at cost to Modesto and Turlock Irrigation for agricultural pumping. Of the 950 MW of the city's peak power, 190 MW is used for municipal facilities, which are largely found in the form of hydropower from the Hetch Hetchy dam. The hydroelectric power is produced by water that flows from the Hetch Hetchy reservoir through two hydroelectric plants, and as water is released from Cherry Lake and Lake Eleanor, two nearby reservoirs.[10]

Biodiesel

According to the San Francisco Department of the Environment, "biodiesel is a domestically produced, renewable fuel manufactured from vegetable oils, animal fats, or recycled restaurant greases. It's safe, biodegradable, and it reduces serious air pollutants like carbon monoxide, particulates, hydrocarbons, and air toxics. Blends of 20% biodiesel with 80% petroleum diesel (B20) can be used in unmodified diesel engines; pure biodiesel (B100) may require engine modifications to avoid maintenance and performance problems.

In its role as a leader in the use of alternative transportation fuels to clean the air, promote renewable energy, and reduce greenhouse gas emissions, San Francisco now has more than 800 alternative fuel vehicles in its fleets. Several city departments and agencies have successfully tested and used biodiesel in pilot programs using B20 or higher biodiesel blends, including the San Francisco Airport, Department of Public Works, San Francisco Municipal Transportation, San Francisco Zoo, and the San Francisco Fire Department."[11]

---

[9] http://www.sfenvironment.org/our_programs/topics.html?ssi=3&ti=6.

[10] http://sfwater.org/mto_main.cfm/MC_ID/12/MSC_ID/145/MTO_ID/344.

[11] http://www.sfenvironment.org/our_programs/interests.html?ssi=6&ti=15&ii=58.

In February 2009, San Francisco became the first city in the United States to begin a pilot program to convert "brown grease" into fuel. The $1.2 million dollar program is funded by state and federal grants in order to open a recycling plant where the conversion processes will take place. It is estimated that 10,000 gallons of dirty grease will be collected per week, which can be converted to about 500 gallons of fuel.[12]

---

[12] Allday, Erin. "SF Converting Gooky Grease to Fuel." San Francisco Chronicle. February 5, 2009.

Chapter 8
Building Sustainability: The Role of K-12 Education

Bill Radulovich

Sustainability – the capacity to indefinitely meet human needs while preserving the environment – is more than just a noble idea. It is quite possibly the central imperative of our time, the Gordian Knot[1] of the 21st century. The impacts of climate change – the political tensions that emanate from competition for energy and other

B. Radulovich (✉)
Project EarthRise, Pleasanton, CA, USA
e-mail: bradulovich@pleasanton.k12.ca.us

[1] As an historical aside, it is of interest to note that, according to Greek lore, when Alexander the Great cut the famous knot in Gordium (in present day Turkey), he freed Zeus's oxcart, the mythical vehicle that symbolized sky and lightening (radiance and electricity).

natural resources, as well as the collision course that looms on the horizon pitting established nations against developing countries – have roots in the *sustainability riddle*. The "riddle" – *how to redirect human behavior toward a critical level of sustainable activities* – is the domain of this publication. The various authors offer a broad array of promising practices, drawing on a wealth of international acumen.

The present author comes to this discussion from a finite but critical perspective – the K-12 education community. In this chapter he will attempt to accomplish two things: (1) to argue that – at least for America – the K-12 education community is key in solving the *sustainability riddle* and (2) to document a case study in which at least part of the riddle's solution is being unraveled.

The argument on behalf of the role of K-12 education in solving the riddle is structurally *elemental*. And the essential elements are the triptych that follows:

Opportunity

In the future, nearly all potential *partisans* in the campaign for sustainability must pass through a common K-12 experience. Schools are ubiquitous: on virtually any given weekday, one-fifth of all Americans are within the walls of a school. Students, teachers, secretaries, custodians, nurses, administrators, and paraprofessionals make up 20% of our national population. With so many gathered in one place, the school setting presents an opportunity that is otherwise absent in American life. From a geo-demographic perspective, nowhere in our culture do we amass so many souls on such a regular basis for such an extended period of time. And perhaps even more important, this experience is America's greatest *equalizer*. By providing all children – regardless of genetic or schematic background – with common experiences over a 13-year period, we have the opportunity to ensure that all individuals partake of *common understandings* – in this case, learning the algorithms that comprise the sustainable living equation. And to make it even more *opportune*, those who graduate from the K-12 experience recycle back into the system (teachers, administrators, etc.) – thus *sustaining and renewing the teaching–learning cycle*. In America, we are yet to take full advantage of this opportunity.

Function

John Dewey, arguably America's most important education reformer, said, "Education is a social process ... not a preparation for life; education is life itself." He went so far as to assert that the purpose of education is "... bringing the child to share in the inherited resources of the race, and to use his own powers for social ends." It is hard to believe that Dewey wrote these words over 110 years ago; such prescience is rather chilling. It is almost as if he envisaged the very notion of *sustainability*!

Similarly, the notion of *educating for citizenship* (Horace Mann, *ca.* 1850) has its roots in American history, and it has traditionally been considered the core purpose for public education, a system indigenous to America. It is thus fitting that sustainability as a curriculum – as well as a value – should be integrated into the American education system. It is wholly consistent with our history as well as the historical narrative that defines the relationship between our political and education systems.

Timing

The natural patterns of human growth and development support the timing of "environmental awareness" during the K-12 years. Behaviors that fall under the aegis of "sustainability" follow from an ethical or moral imperative. One chooses to behave "sustainably" by consciously acting in prescribed ways – composting, recycling, commuting via bicycle, and turning off lights are a few such examples. In each case, the individual's motivation for making the behavioral choice is primed by socially driven (and rewarded) norms. Parallel processes are imbedded throughout every child's early years; these processes are the grist of socialization, and they perpetuate the basic cultural conventions of society.

This aspect of human growth is generally referred to by psychologists as the "stages of moral development," and much of what is theorized about these developmental stages is attributed to the American psychologist Lawrence Kohlberg (Harvard Center for Moral Development). Kohlberg partitions moral growth and development into six distinct but overlapping stages. Beginning with the pre-conventional stage (self-interest/punishment avoidance), children pass through increasingly more globally aware stages (conventional and then post-conventional stages). With the development of each stage, the individual increasingly moves from self-interest to the understanding of interdependence, social reciprocity, and universal justice. According to Kohlberg and others, the developmental process is *constructive*: it follows a linear sequence, and each of the six phases precludes its successor. As such, care must be exercised at each juncture in order to build on the next. The quality of the outcome rests on the quality of attention devoted to each stage. Kohlberg, it is of interest to point out, postulates that only about 25% reach stage six, while the majority of individuals stall at stage four.

Given that the K-12 experience presents itself during the period in which the foundation for moral and ethical formation is largely constructed, it presents the ideal time in which to impart the principles of sustainability. Nature's "developmental window of opportunity" presents itself during the fertile K-12 years. Failure to leverage this window – to ignore the calendar that governs human development – will surely inhibit, if not undermine, the transformation of our society into one that is sustainable. Additionally, it may lead to inaction on a host of other critical global issues, like war, poverty, and pandemic.

This triptych – *opportunity, function, and timing* – are three practical and strategic reasons for emphasizing the teaching of sustainable principles and practices in the K-12 curriculum and as part of the greater school experience. Because the K-12 setting creates the ideal capacity for doing so, it presents our best *opportunity*. Because it is thoroughly consistent with our history – the cultural, social, and political record – it presents a vital *function* for our democracy. Because it embraces the theory and science of human development, the K-12 years provide the ideal *time* in which to present these principles and practices. By understanding the elemental relationships between the elements of this triptych, we can better leverage and control future outcomes. If climate science is to be believed, we cannot overlook this opportunity.

Case Study: Pleasanton Unified School District

Today there are abundant examples to suggest that children are developing *environmental awareness* within America's K-12 education system. In still others, environmental precepts – fundamental maxims based on ecological principals – help transform awareness into behavior. Awareness precedes action, but does not ensure it. *Motivation* must exist in order to *operationalize* thought and cause the initiation of action.

The example that follows has been selected because it represents one such case. Others exist, and by the time this publication reaches its intended audience, it is the author's hope that the present case study will have become commonplace by contrast to those that have emerged since. In the event that his optimism is overstated, he offers the story of Pleasanton School District.

Pleasanton is a suburban school district of approximately 15,000 students and is located in the San Francisco Bay Area. It is surrounded by ridgelines that separate it from both the San Francisco Bay and California's Central Valley. With a population of 64,000, Pleasanton's demographic profile is largely upper middle class, while its ratio of Caucasians to non-Caucasians is 50% higher than the national average. The reported "attained education" level of Pleasanton's typical adult is 16 years, just short of college graduation.

Pleasanton has a rich history, but none of it is steeped in factors that would make it uniquely suited for taking a groundbreaking role in enacting sustainable practices. The city began under a land grant (Mexico) and for approximately 100 years was the site of large ranchos. Following California's admission into the Union of States, wealthy San Franciscans, like William Randolph Hearst, built weekend homes in the hills of Pleasanton. Later, the town enjoyed a brief film industry association (1919–1943). Growth came somewhat slowly to Pleasanton until the 1980s when, because of its position at the juncture of two major state highways, it became the site of one of the largest business parks in America.

With the rapid growth that followed during the subsequent 20-plus years, residents appear to have developed a degree of "environmental circumspection." At both city council meetings and the polling places, concerns about development on the verdant ridge lands that bound its western border were demonstrated. If any precondition was created that might account for Pleasanton's fast-track journey down the "green highway," it may stem from the frequent "stops" in which Pleasanton residents were forced to halt at the intersection where *growth* crossed *nature* during that period. The thoughts generated, and the dialogues that developed, while paused at these crossroads may have seeded what followed.

What happened in Pleasanton is not different from what is happening in many communities around the planet, something suggested throughout this book. The difference is not so much one of kind; rather it is one of *degree*.

During the winter of 2005–2006, events began to take shape in Pleasanton that historically effectuated the *Green Record*.[2] The events referred to here occurred primarily within the city's school system, Pleasanton Unified School District (PUSD). These events were predated by yet another series of events, which in the aggregate, resulted in PUSD becoming the first "solar school district" of scale[3] in California, as well as one of the first in the USA. Perhaps more important than this achievement, however, is the *means* by which it was achieved. Out of these circumstances, a unique financial arrangement – a "solar power purchase agreement" (SPPA) – was developed and successfully executed. As you will read later in this chapter, the SPPA has promise for our schools, our environment, and our economy. In the pages that follow, the author will track the events and conditions that preceded and succeeded the SPPA, which was spawned in PUSD.

Dating back to the 1970s, PUSD schools participated in activities related to horticulture. These were scattered efforts, largely related to school clubs (Future Farmers, 4-H, etc.). Whether the teacher/advisor of these clubs infused environmental/ecological education concepts was coincidental, depending on the teacher. By the 1990s, these clubs became nonexistent for lack of interest. However, the concepts of organic gardening, composting, and water conservation began to find their way informally into one school site. Parents, along with supportive teachers and administrators, created a small working garden at an elementary school site, Walnut Grove. Again, these were informal structures – a product and function of the persons who operated them and not part of an adopted curriculum or identified educational objective. They did, however, set a fertile stage for what followed.

The teaching of composting, largely an optional recess activity for interested students, led to a recycling program. The composted materials, primarily paper from classrooms, outgrew their bins; the schools simply had too much paper waste to manage in the small composting area. Recycling became the answer for this problem.

At one of the elementary school campuses, Walnut Grove Elementary, a third-grade teacher noted the issue. She researched local recycling companies and was pleased to find that the local waste manager, Pleasanton Garbage, was supportive of such an idea.

---

[2] The concept of "green record" has evolved from a generalization used by environmentalists in assessing various associations (political, social, and economic) to an industry standard in marketing. Today, marketing and public relations firms (e.g., AMP Agency) calibrate public sentiment for their clients and advise on how they may best affect their Green Record.

[3] A small, single-school 20 kW system was built north of San Francisco some months earlier.

The garbage company was among the early enlistees into the "business" of recycling, largely due to the city's rapid growth and the dwindling landfill capacity within the city. Efforts begun by the company in 1991 resulted in a 50% diversion rate from solid landfill by 2003.

Partnering with the garbage company, Walnut Grove began "The Gobblers" program. The Gobblers were 30-wheeled 60-gallon waste cans made of recycled plastic. These receptacles are stationed outside of classrooms throughout the week. Classroom paper waste is deposited at the end of each day into each classroom's respective Gobbler. On Wednesday of each week, third-grade students collect the Gobblers and station them curbside in front of the school for pick-up/recycling. On average, Walnut Grove diverts 76,000 pounds of paper from the solid waste stream yearly.

For many students this is their introduction into "service learning" – learning "life" concepts through service to others. The Gobbler program quickly spread from Walnut Grove to other PUSD schools, and it is now a district-wide undertaking. In 2007 over 212,000 pounds of paper was recycled by PUSD schools.

The "greening" of PUSD was accelerated significantly in 2002. At that time, a community member and Walnut Grove PTA president developed an idea for promoting environmental practices at schools, the Go Green Initiative (GGI), which began as a sheaf of notes on a kitchen table and has since then flourished into "the largest environmental education program in the world" (GGI press release, December 2008). Go Green has 1.5 million student members and 113,000 member teachers, and it operates in 13 countries.

GGI began on Pleasanton's Walnut Grove campus in 2002. At that time, Jill Buck, GGI founder and executive director, approached the Walnut Grove Principal Bill Radulovich with the concept. Since it was an environmental-friendly campus – with gardening, composting, and recycling already in practice – her proposal was openly embraced.

As the program gained traction on the elementary school campus, Mrs. Buck went to work to build its presence. Soon, all 15 PUSD schools were operating under the banner of the GGI. Most began with implementing recycling programs, while some included composting and reuse practices. Mrs. Buck wrote an initiative for the California State PTA (GGI), which was presented at the association's convention in 2003. GGI is now in all 50 states and has been adopted by eight state PTA councils.

The GGI helped PUSD to coordinate efforts with waste management agencies. It also assisted teachers in receiving minigrants for environmental projects. Schools participating in GGI were required to develop site leadership teams composed of parent/community and staff representatives. Most schools formed student clubs as well. At Walnut Grove Elementary, two clubs developed: Roots and Shoots and the Ecology Club. These clubs provided opportunities for students to learn about environmental stewardship (Ecology Club, grades K-5) and then to practice stewardship (Roots and Shoots, grades 4/5). The practice of stewardship is expressed in many ways: painting caveats onto storm drains that lead to the ocean/bay habitats, creating kiosks for dog waste in which recycled bags are put to use, cleaning up local habitats, and raising money to purchase/protect tropical rainforests. With regard to the latter, the students of Walnut Grove have "protected" over 300 acres of tropical rainforest (Riff Valley, Borneo, Central America) since 2005. In 2003 the city of Pleasanton adopted the GGI, becoming the first city to do so – and thereby subscribing to a common platform with the education community.

The success of GGI, both on its home turf in Pleasanton and abroad, is that it provides a planning guide for participating schools – one that can be modified to any situation. CGI is not a prescription: schools may choose to participate based on their level of readiness, and no restrictive preconditions for membership exist. The GGI organization records and reports aggregated data from participants. Any school that is inclined is given an opportunity (taking license with the words of Earth First's David Brower) "to act locally" while participating globally.

To provide member schools a sense of being part of a larger effort, GGI posts data, like the following numbers for cumulative savings of resources (equivalents) since January 2006:

- Saved 208 billion BTUs of energy use
- Avoided 15,000 metric tons of greenhouse gas emissions
- Saved 1.3 billion gallons of water (paper recycling alone)
- Conserved the equivalent of 1 million gallons of oil
- Prevented 64,000 cubic yards of landfill space, due to paper recycling alone

In the "greening" of PUSD, GGI played an important role. During the spring of 2004, the school board adopted a strategic plan that was developed by a committee (51 staff and community members) over a 14-month period. One of the eight "core business" pillars for PUSD – Strategic Plan Goal 3 – was identified as "environmental awareness." The plan, predicated on "futures forecasting" – planning on the basis of how the "future is most likely to present itself" – is designed to prepare students for the world that they are most likely to encounter in the years ahead.

As a result of the strategic plan, environmental education shifted from a de facto ancillary practice to a de jure component of the K-12 curriculum. In addition, a standing committee, the Environmental Awareness Committee, was formed to take on the role of championing and overseeing the implementation of the strategic goal. The committee is comprised of educators, parents, students, business representatives, and interested community members. They report to the general public through a web site as well as yearly broadcast and published school board reports.

As Pleasanton's "green movement" spread from the garden at Walnut Grove Elementary School to the other 14 schools, additional programs sprouted, and PUSD's public profile grew as well. With this newly gained status, Walnut Grove applied to the local utility company, PG&E, to become one of its first "Solar Schools." It is important to note that the "solar" part of the "Solar Schools Program" is more symbolic than actual. This is not to diminish the value of the program. The term "solar" refers to the *learning* about solar technology rather than actually developing a working system for renewable energy.

The Solar Schools Program was initiated by the California Energy Commission in 2004 and is administered by California utility companies: PG&E, Southern

California Edison, and San Diego Gas and Electric. The program was funded by the California Energy Commission through its Renewable Energy Program (ERP) along with matching funds from the California Attorney General's Alternative Energy Retrofit Account (AGAERA).

"Solar Schools" were provided with two important forms of support: a 1 kW photovoltaic unit and teacher training. The 1 kW unit, known as "PV on a stick," is actually just that – a PV panel resting atop a steel pole. The PV units, unlike those in commercial and residential use, occupy auspicious places on their respective campuses.These units produce barely enough energy to power one classroom; however, they serve as a bigger-than-life symbol. In conjunction with the curriculum and teacher training, they provide a powerful opportunity to make a very abstract concept – renewable energy – accessible to young children.

The Solar Schools Grant included training. The National Energy Education Development (NEED) program was contracted by the Energy Commission to provide this component. NEED, headquartered in the Washington DC area, hired a California staff to work with granted schools. Walnut Grove began with summer training for its science specialist. She spent 2 days in Sacramento, receiving training and hands-on resources for teaching the curriculum. The curriculum, which begins with the general concept of "energy," expands into its various forms, with an emphasis on solar.

The staff at Walnut Grove was impressed with the quality of the solar program. Students in third grade became the targeted population for introducing the curriculum. The initial unit of solar instruction culminated with solar projects. Students designed solar-powered cars, used photosensitive chemicals to do solar art, cooked on solar stoves, and even defined a functioning "Solar Barbie House." NEED, PG&E, and the California Energy Commission joined the Walnut Grove community in September 2004 for "Solar Fiesta" – a celebration of the newly formed partnership. Nearly 1,000 attended, including local politicians, news media, corporate executives, and other interested educators.

The interest in the local education community spread. By December 2004, NEED was presenting full-day weekend workshops for Pleasanton teachers. Individual participants received, along with the training, over $1,200 worth of teaching materials, a small stipend, and a catered lunch. These workshops inspired new ideas about the teaching of science in general and in teaching about energy in particular.

The "new" environmental curriculum has taken a significant departure from the practices of the past. It is data driven and founded on scientific principals. The specie-specific and habitat-specific programs of the 1980s and 1990s have been replaced with programs that offer a more *analytic* approach. As such, they teach from a curriculum that goes from *global concepts* (energy, matter, systems in nature) to specifics – like individual species and habitats. The "old practice" – gardens, composting, recycling, and reusing – all persist. They just have taken on new, enhanced meaning in an ever-expanding learning experience.

The Walnut Grove principal was among those inspired by the ideas generated in the NEED workshops. While agreeing with the value of *teaching* the facts and concepts surrounding energy, he postulated that the *practice of implementing them* was the next logical step in the "greening" process. Toward that end, he drafted a proposal – which he submitted to both his district CFO and PG&E for funding – that created the "eCoach Program."

The eCoach program is currently in its second year in PUSD. It enjoins staff and students in the mission of saving energy. The program is organized as follows:

- Each school is asked to provide an eCoach;
- eCoaches are provided the equivalent of a coaching stipend (e.g., high school sports coach) for their efforts;
- All eCoaches go through yearly training programs provided by NEED and PEAK/Energy Coalition;
- eCoaches develop eClub or eTeam on their respective campuses;
- Each club develops strategies for reducing energy waste/plug load on their campus, based on the individual culture of each school;
- The eCoach coordinator, along with an independent energy consultant, develops energy-usage baselines for each school;
- When schools reduce monthly kilowatt-hour consumption below the baseline, they are rebated (eBates) the savings;
- Each school determines how its eBate is to be used;
- eCoaches meet monthly to review their progress, share successes, and build on emerging idea.

The eCoach program was immediately successful. Coaches stepped forward from all schools. NEED provided initial workshops and equipped the eCoaches with the tools and training to conduct energy audits at their schools. Three months after their initial meeting, the eCoach program posted up its first 1-month performance data – resulting in a single-month reduction in the district's utility bill of over $9,400. Currently, the program saves the district over $130,000 per year in utility bills. In September 2007, because of state-funding cuts, the eCoaches voted unanimously to temporarily forego the eBates and to "re-rebate" them back into the district general fund until the district/state recover from a fiscal slump.

The eCoach program looks somewhat different at each campus. They each reflect the passion that drives their respective coach: where some coaches have a passion for composting, others love gardening, while others leverage technology. Still others mine the promising vein of "green culture." This involves integrating art, music, writing, and social activities into the "green life."

"Green Culture" has been a very productive enterprise at Walnut Grove School, where environmental murals, gardens, stage productions, technologies, and literature are ubiquitous features. Four large murals adorn the campus, all commemorating the wonders of nature. There are three major gardens on the campus. One produces seasonal organic vegetables, herbs, fruits, and flowers. Monthly, the produce from this garden is featured as "Harvest of the Month" item. The school food service workers collaborate with parents to prepare special-tasting items (e.g., spinach salad) for students to sample during lunch. Another, the Peace Garden, is maintained by the Ecology Club. This is a refuge for wild birds, providing plants that attract insects that are preferred by Pleasanton's feathered population. It also includes birdbaths, as well as seed posts and liquid feeders for hummingbirds. The California Native Specie Garden is a study area. There, students care for and observe indigenous plants, including several herbs that figure significantly into the life and lore of Pleasanton's indigenous people, the Ohlone Indians (fourth-grade social studies curriculum).

Perhaps the most powerful garden is the K-Garden. Literally a "kinder garden," this is a 120-foot-long strip of earth that divides buildings that house kindergarten students on one side and first graders on the other. Each year, on the first day of spring, the soil in this strip is cleared and tilled for seeding. In ceremonial form, the students are all gathered around the strip, where they sing a series of songs about spring, flowers, and nature. Each is given a handful of wildflower seeds. On cue, each casts his/her seeds onto the earth, where parents and staff rake them into the soil. As the seeds sprout and grow, the students observe, measure, and record their development. Measurement, graphing, statistics, data analysis, and basic mathematics are naturally and meaningfully woven into the curriculum. But the most powerful – and intended – value of this garden is neither curricular nor cognitive. It falls into the affective domain: young children develop a kinship with the Earth. Just a matter of a few feet from their classroom door, they are met with the wonder of life unfolding before their eyes. At least twice daily they are bathed in a vision of beauty – the brilliant color of over 100,000 wildflowers is their last vision as they enter the classroom and the first as they exit for home. The staff at Walnut Grove has come to recognize that "environmental stewardship" does not take root out of fear (global warming) or sentiment (animal extinction). Rather, the determination to protect all living things on the planet begins by falling in love with its natural beauty. That's the underlying purpose and role of the "Kindergarten" at Walnut Grove.

A special area at the end of the strip is reserved for sunflowers. This is because the "PV on a Stick" stands at the end of the strip, overlooking the entire garden area. The sunflowers, nature's heliotropes, grow next to man's heliotrope – the PV on a stick. The symbolism is powerful, but the conversations that ensue – relating how light can be converted to chemical energy (photosynthesis) or electricity (engineered PV), or how both work toward the same end (reducing carbon in the atmosphere) – are all rich learning opportunities.

The environment "takes stage" when special presentations take place at the school. Such presentations range from visits by a John Muir impersonator to lessons about the water management delivered by a comedian on a unicycle to environmental rock bands like Scientific Jam and the Banana Slug Band.

When the daily aspects of school life are integrated fluidly with environmental concepts, the net impact is *cultural* – the symbolic structures that give human activity significance and meaning are altered and reshaped. Norms, values, and behaviors are shifted by learning that is at once multidimensional and dynamic.

The changes in PUSD's cultural landscape that took place between 2002 and 2005 must have been noticed outside of the community. They set the stage for the next, and perhaps most important, event in the greening of PUSD.

In September 2006, the District Environmental Awareness Committee reconvened for its third year. In attendance was a representative from one of America's largest oil companies. She had come because her company had recently developed a new division, a renewable energy group. As a member of that division – and an environmental activist during her pre-corporate years – she volunteered to participate as a member of the PUSD Environmental Awareness Committee. During a break at this particular meeting, the principal of Walnut Grove introduced himself to the new committee member. Business cards were exchanged and – unknown to either at the time – events were set into motion that would transform PUSD, add an historical footnote to the Green Record, and pave the way for others.

Emails followed the card exchange, leading to a face-to-face meeting in October 2006. At the meeting, the two – principal and junior executive – discussed solar energy and the possibility of bringing a large-scale solar installation to Walnut Grove School. After a follow-up meeting, both presented the idea to their superiors. In both cases, the idea was warmly embraced.

The next meeting involved the principal and the junior executive's director. At that meeting the oil company executive suggested expanding the initial idea – the creation of a solar school district.

At this point the school district business and maintenance leaders were brought into direct discussions with corporate technical, administrative, and financial personnel. A photographic flyover was arranged in order for the technical staff from the oil company to scope the parameters of the job: plans were drawn, calculations were carefully drawn, and a detailed proposal was delivered to the PUSD administration.

While the proposal was impressive, in terms of both its detail as well as its design, the school team – CFO, maintenance director, and Walnut Grove principal – knew that they were on unfamiliar ground. Technical issues were raised that surpassed their experience and expertise. Financials were uncertain – generally untested altogether in an education setting. Lacking historical precedence, liabilities were difficult to identify, much less address. While the idea of being a "green solar school district" was alluring, the idea of putting large amounts of public capital at risk was too daunting to move forward without further consultation and diligence. PUSD asked the potential SSP to put discussions on hold until they addressed these concerns.

To guide the district through these issues, a local energy management consultant was hired. Luckily, he had a strong background and interest in renewable energy – as well as a vast network of associations.

Upon reviewing the proposal from the oil company, he suggested that he "float" the concept within his circle of industry associates. No RFP was tendered. No formal bid specs were drafted for sourcing. What happened next proved that PUSD had well prepared itself for a "ripe" moment.

Within days after putting the initial oil company proposal on hold, another corporate giant (also a Dow 30 Industrial) began courting PUSD. This corporation, a multinational engineering firm, will be referred to as "Suitor #2".

Suitor #2 arrived one afternoon in mid-November in the form of a team of well-dressed marketing, technical, and operations executives. They had not had time enough to develop a written proposal, so they presented their best oral presentation – a live conference call among the PUSD team, Suitor #2's staff, and a senior VP who was on vacation with family in the Cayman Island.

The visiting team and their vacationing boss provided a sketchy, but compelling, proposal – good enough to cause PUSD to tell the oil company to continue to wait as the district pursued a further course of due diligence. In the days that followed, the PUSD team awaited the formal, written proposal from Suitor #2.

While PUSD was weighing the two competing proposals, another "suitor" entered the picture. This one, also (at that time) a member of the prestigious Dow 30 Industrials, was Honeywell Business Solutions, a division of Honeywell Corporation. The Honeywell proposal differed significantly from the first two proposals – offering the most attractive solution for PUSD.

This solution involved what was refereed to at the time as a power purchase agreement (PPA). Currently, to distinguish this agreement from those involving other energy forms, they are commonly called SPPAs (solar power purchase agreements), the designation that I will use for the agreement henceforth.

The PUSD agreement represented a unique approach to an otherwise intractable problem. Utility bills represent the second largest expense for schools, exceeded only by teacher salaries, which typically represent 80–90% of their operating budgets. The remaining dollars do not allow for the capital outlay that would be needed to finance and build renewable energy systems. The Federal Tax Code (IRS 3468) and the recently reinstated Federal Incentive Tax Credit Bill, which provide a 30% rebate for installing solar panel, do not apply to schools – adding yet another obstacle for schools to become sustainable entities.

The Honeywell SPPA provided PUSD with a solution that helped overcome the structural, financial, and logistical issues that would otherwise cause a school district to balk. First of all, it required no front-end capital outlay on behalf of the district. Next, the district would not be responsible for maintaining the system. Finally, the arrangement would lower the district's operating costs and fix energy costs at a below-market rate for 20 years.

The SPPA designates Honeywell as the solar service provider (SSP). As SSP, Honeywell assumed responsibility for designing and building the solar power generation system. SPG Solar, a Northern California-based solar energy contractor, served as the solar integrator – constructing the system for Honeywell. PUSD, the *customer*, is actually the "host" of the system, providing Honeywell (SSP) with easements that allow them as SSP to have access to school sites as well as the usage of the architectural systems (electrical and mechanical) that support the solar infrastructure.

The arrangement made Honeywell a "secondary utility company" for PUSD, supplying approximately 20% of all of PUSD's electricity needs. According to Honeywell's press release (June 1, 2007), this agreement was a "first ... where the contractor acts as the utility for a K-12 school district."

As the SSP, Honeywell received the Incentive Tax Credits, valued at $2.4 million. They also received Renewable Energy Credits (RECs). These "credits" are a tradable commodity, one in which the "renewable" value of the solar energy (or other renewable form) is monetized.

At the time of the SPPA's conception, Renewable Energy Credits did not have a well-established value in the USA. Today, RECs have become a valued commodity and are being traded on several competing carbon exchanges in the USA. The Kyoto Protocol (1997) generated a host of international carbon exchanges nearly a decade back.

As per the SPPA, PUSD can purchase all of the energy that is produced by Honeywell. The cost was set at 11.4 cents per kWh with a 4% automatic annual *accelerator* (increase) from that rate base. This rate represents a 30% below-market cost per kilowatt-hour. This translates to a yearly savings of an excess of $130,000 to the school district. With most of California's power plants using fossil fuel to spin their turbines, the cost of kilowatt-hour is expected to escalate at double-digit figures in the coming year. The low starting rate, combined with the modest *accelerator*, promises millions of dollars of savings for PUSD throughout the term of the SPPA.

The "genius" in this approach is that the "financing" of the system is transformed from a note of indebtedness into the monthly utility bill. Instead of having to find third-party financing for the project, it pays for itself while paying for monthly utility costs. What is even more remarkable is that the transformation of potential debt to *anticipated operating cost* results in lower utility bills. Green energy that pays for itself – even while creating a net reduction in the debit side of the ledger – is almost too good to be true. But it is, and PUSD is a living proof.

The agreement expires in 2027. At that time, PUSD will have the option to take ownership of the system for a nominal fee. PUSD will then become *its own utility* and will sell energy back to the ISO (grid) on weekends, summers, holidays, and daily peak hours (school energy demand falls off significantly after 4 p.m.). Honeywell has an SPPA obligation to continue maintaining and insuring the functioning of the system for another 10 years, until 2037.

The PUSD solar array is spread over the rooftops at seven different schools. Totaling nearly 600 kW in production, the size of the system was determined by four primary factors: (1) the availability of rooftop space on which to locate the panels, (2) the orientation of the district buildings with respect to the sun's arc, (3) the need to alter infrastructure in order to make the project costs and benefits come into balance, and (4) the anticipated permanence of the buildings on which the panels were to be placed.

With the completion of the solar program in Pleasanton, a celebration was held, "eFair Pleasanton." This event followed Earth Day and was staged to celebrate the corporate, government, and nonprofit partners who helped shepherd the project to fruition. A press conference was held prior to the fair, with representatives from the district, PG&E, Honeywell, and SPG pulling a ceremonial switch to signify the "solarization of the school district."

The highlight of the event, however, involved a very unlikely crew – a dance troupe from Liberia, Africa. The Liberians were at "eFair" to thank Pleasanton's Walnut Grove School for reaching out to their school and village, Johnsonville.

Walnut Grove's student Club, Hands Across the Water (HAW), had been providing support to the students of Johnsonville since 2002, after civil war had decimated it and cut off energy to the city.

Lacking electricity, the school stopped to function. Money was raised by HAW and generators were sent. The cost and availability of gasoline and kerosene, however, soon caused the generator to stop operating. With no available fuel for cooking or heating, the people of Johnsonville began to deforest their surroundings. That's when the HAW students, along with the school counselor and principal, set about finding a solar solution: solar ovens and portable solar generators. The students raised funds and ovens, and solar generator/battery devices were sent. So were books, clothing, and cash – along with cards and words of support.

The dance troupe, most of whom were on tour in nearby Oakland, came to eFair Pleasanton to express their appreciation to the HAW kids. At the height of the event, the Liberians took the stage, drums beating, bodies shaking, and voices singing. The auditorium was electrified by effect – and the host citizens of Pleasanton were visibly moved, joining in by improvising their own dance steps on the floor of eFair.

Elementary-age students, reaching across the globe with their knowledge of green tech, saw that they had made a difference. Something very special happened that night – and like so many other events in the "greening of Pleasanton," it happened very naturally, almost serendipitously.

The process that led to this moment engendered many other enhancements. The Energy Coalition/PEAK, a nonprofit based in Irvine, California, came forth to provide additional staff development and training. They worked with science teachers at the fourth and seventh grades to identify units of study from the California State Content Standards that could be replaced by a "supercharged" energy curriculum. This curriculum included hands-on, experiential learning opportunities that better demonstrate the core relationships that gird the concept of sustainability. Along with the training came curriculum materials and media. They provided school assemblies, in conjunction with PG&E, and they helped the Pleasanton schools to "swap" over 10,000 CFLs (compact florescent lights) for incandescent bulbs in Pleasanton households – all for free.

Other corporate organizations took note of the forward-thinking strategies taking place within the schools and offered to partner with PUSD. Two such companies – one an energy management and software company and the other a cellular communication company – teamed up to offer PUSD a world-class energy management system (EMS). This system utilizes "smart meters" and "telemetering" to provide detailed real-time feedback about the energy consumption patterns within various buildings. The smart meter is an electrical meter that replaces conventional utility grade meters. It has the capacity to break down data into detailed reports. The reporting is then sent via telemetering – using cellular technology to transmit the data to computers and cell phones.

The software built into these systems allows the user to set parameters for the reports that are generated. For example, if a building in Walnut Grove Elementary School is unexpectedly consuming an abnormally large number of kilowatts, a message would arrive on the custodian's and the principal's emails. If the trajectory for that day's use appears as though it may threaten to reach "peak demand," a text message would be sent to their cell phones, along with the cell phones of district-level operations managers.

This would allow the problem to be remedied quickly, but it also allows the school district to negotiate an enhanced rate structure with the primary utility company, PG&E. By having such immediate and accurate information about energy consumption, the district is able to create a new rate structure that lowers their regular kilowatt-hour rate but raises the penalties for reaching peak demand – thus providing an "insurance policy" to PG&E. Utility companies typically add a margin onto the rates for commercial consumers in order control for the costs of operating during brownouts or blackouts. California experienced rolling blackouts in the early 2000s and again in August 2005; Texas had similar issues in April 2006. By installing a system that protects the user from reaching peak demand, the user can negotiate a rate that removes or reduces the margins that are structured into the rates paid by other public entities.

This system, also a *first* in a school setting, is being installed in beta facilities as this book goes to print. At the same time, other potential corporate partners are stepping forward to share the cutting edge with PUSD. One such corporation is a worldwide leader in the management of commercial real estate. With an ambitious "green agenda," this US-based company has approached PUSD with the desire to "explore potential partnerships." Another organization, a Euro–American association that represents "green partners" seeking to import European green-tech concepts into the American workplace, has presented itself as a potential partner with PUSD in exploring the application of these concepts in the school setting.

Corporate "partners" are generally empiricists by nature. They rely on hard data, not sentimental narrative. That, among many reasons, is why they are able to successfully navigate in a competitive business climate. So it makes sense that they apply similar standards to their public affiliates.

The "greening of Pleasanton" has yielded some powerful performance data. In October 2008, Pleasanton's Walnut Grove School was named "School of the Year" at the Fourth International Earth Summit in New York – an international recognition of the school's achievement.

In the 6 months leading up to November 2008 alone, the Pleasanton schools have – by virtue of the programs described in this chapter – contributed to *shared global well-being* in the following ways:

- Offset more than 740,000,000 tons of carbon dioxide through PV;
- Created enough clean, green energy to offset over 1,000,000 miles of automobile use;
- Saved the equivalent of 42,000,000,000 BTUs of energy through recycling;
- Conserved 280,000,000 gallons of water through recycling.

Pleasanton's history, as mentioned earlier in this chapter, is rich. It continues to become richer through innovation and determination. In particular, the combined efforts of a supportive school board, a progressive citizenry, an informed electorate, committed teachers, strong leadership, effective community-based marketing, and fortuitous timing all add up to the prime conditions for continued improvement. More important perhaps, a legacy will be handed down that inspires other communities to board *the fast track* on *the green highway.*

One aspect of this legacy is the formation of an NGO (nongovernmental organization/nonprofit) that will work with schools throughout the state to replicate the benefits that are enjoyed by PUSD. Such efforts hold great promise. If, for example, all 9,000 California schools were to adopt/adapt the Pleasanton-style green design, there would be a yearly reduction in utility costs of over $472,000,000.

On a national level, it is difficult to compute precise numbers. However, it is relatively safe to extrapolate a yearly savings close to $10 billion per year. These dollars represent "the conversion of kilowatts into curriculum" – the opportunity for the educational community to redirect large amounts of capital to where it belongs, the classrooms. It also represents the transformation of carbon-dirty kilowatts into green, sustainable energy.

Using the Pleasanton model as a basis, the "greening of American schools" would directly generate over $200 billion dollars into the economy, creating 6.8 million jobs, each generating more dollars into the economy. At the time in American history in which this book is being written, economic stimulation and new job creation are sorely needed.

Conclusion

Perhaps some useful lessons can be harvested from PUSD's "green history." While the Pleasanton story is but one relatively small case study, it presents a successful model that can be replicated, improved upon, and may even be inspirational.

Robert Collier, one of America's earliest and well-respected motivational authors, wrote "Success is the sum of small efforts, repeated . . ." This case study is presented to suggest that the model – or similar models – might be scaled up to create large-scale solution(s).

The present writer believes that the *sustainability riddle* – how to sufficiently change human behavior so as to reverse the effects of climate change – can be solved only if it is "sent to school." Specifically, he suggests sending it to *all of America's K-12 schools*!

At this point in time, the triptych of "essential elements" necessary for this change – *opportunity, function, and timing* – is very much in evidence in America. We have capacity; we have workable blueprints. Now we need to combine them to produce the *citizen workforce* needed.

As stated in the chapter's opening, we face the Gordian Knot of the 21st century. Just as Alexander seized the moment in cutting his knot in Gordium 2,500 years ago, the moment for ours is at hand.

Chapter 9
Google's Clean Energy 2030 Plan: Why It Matters

Thomas Jensen and David Schoenberg

T. Jensen (✉)
Enterprise Futures Network, San Francisco, CA, USA
e-mail: tjensen@enterprisefutures.net

W.W. Clark II (ed.), *Sustainable Communities*, DOI 10.1007/978-1-4419-0219-1_9,
© Springer Science+Business Media, LLC 2010

Introduction

In late 2008, Google introduced its Clean Energy 2030 plan. It outlined a "potential path to weaning the U.S. off coal and oil for electricity generation by 2030 . . . and cutting oil use for cars by 44%." This chapter provides an overview of Google's plan, and a rationale as to why a company that is generally known as a leader in Internet and information technology is weighing in on this national debate and commentary.

Google's initiative is a departure for profit-making entities. To date, most broad, economy-wide clean energy proposals have come from nonprofits like colleges and universities and the government sectors. Their energy plan and actions to promote it are both new and unique and illustrate how a private company can integrate energy and environment advocacy into bussiness strategy. Few companies are doing that today!

Google's Relevancy in the Renewable Energy Arena

Why does Google's Clean Energy 2030 plan matter? After all, Google is a 11-year-old company commonly known for its Internet search engine and various online applications such as Gmail or Google Maps.

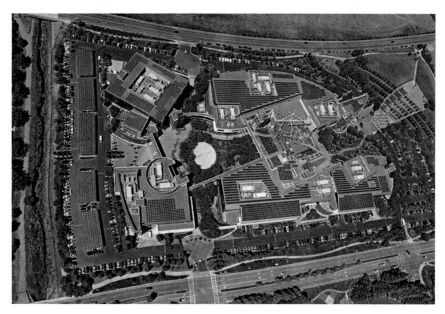

Fig. 9.1 Google campus. The famed "Googleplex," the company's headquarters from above – with solar panels on nearly every building

Corporate Social Responsibility and Google as an Energy User

The Clean Energy 2030 plan is part of Google's overall strategy to operate an environmentally friendly corporation locally and globally while seeking ways to lower energy costs. The company has made a commitment to go carbon neutral and is also working on applications, such as PowerMeter, which is geared toward enabling users to go green.

Energy Initiatives Under Google.org

Clean Energy 2030 was introduced under the Google.org umbrella, a structure the company calls a "hybrid philanthropy" that can draw on substantial financial resources: 1% of Google's total equity (equaling around $148 billion in August 2009) as well as 1% of the company's profits (total profits in the second quarter of 2009 totaled $1.5 billion).

Google.org pursues three main initiatives: energy, the environment, and global poverty. Its energy initiatives generally fall into two categories: the RE<C and RechargeIT initiatives.

RE<C is the shorthand formula that immediately defines this initiative's goal: to make renewable energy (RE) more affordable than coal (C) "with a goal of producing one gigawatt of renewable energy capacity." Three key strategies are employed toward reaching this goal. Primarily, it "is making strategic investments and grants." These investments are externally supporting such companies as Makani Power and BrightSource as well as internal activities in clean energy through Google's newly created renewable energy R&D group. In total, Google.org has invested over $35 million in outside institutions under the RE<C umbrella alone. Additionally, Google.org pledges "to continue to advance public policies that accelerate the development of renewable energy." Finally, it plans to use its products and unique resources to "unlock information that enables innovation and raises awareness about the benefits of renewable energy."

RechargeIT is Google.org's initiative to promote the use of plug-in vehicles in order to reduce "CO_2 emissions, cut oil use, and stabilize the electrical grid." Similar to RE<C, the RechargeIT program has invested substantial sums in external research (nearly $4 million total), supported public policy initiatives, and, most notably, launched its RechargeIT Driving Experiment, centered around a headquarters-based fleet of plug-in hybrids that are part of the company's employee car share program. Through the Experiment, RechargeIT collects extensive data and publishes it on the *RechargeIT web site* as well as the accompanying *blog*.

Google and GE Partnership

Despite nearly $50 million dollars invested in external entities, the mid-September 2008 announcement of a partnership between Google and General Electric created

Fig. 9.2 Google founders Larry Page and Sergey Brin charging one of the RechargeIT plug-in hybrids

the biggest headlines yet. The two technology power houses will first collaborate in Washington, D.C., on mounting "a major policy effort to enable large-scale deployment of renewable energy generation in the United States." Second, they are looking to develop and deploy "smart" grid technology. And finally, the two companies will work together on advanced technology projects around plug-in vehicles and enhanced geothermal systems.

Considering the number and scale of all these efforts, one can conclude that Google has become an important player in the renewable energy arena, and its Clean Energy 2030 plan should be taken seriously.

Overview of Google's Plan

The overall goal of Google's Clean Energy 2030 plan is to propose a roadmap on how to transform the American economy from "one running on fossil fuels to one largely based on clean energy." The plan goes into considerable detail, laying out the specifics of how much fossil fuel-based electricity generation would be reduced by (88%), the declining effect on US CO_2 emissions (49%), and how many gigawatts (GW) would be produced by various renewable electricity sources. Most notable among the plan's projections is perhaps the economic impact of the plan with a financial savings of $820 billion over 22 years.

The detail with which the plan is presented is also gaining recognition in the academic world. "Google has put together an extremely useful piece of analysis. Many people talk about drastic reductions in [green house gases], but Google has provided a description of what a low carbon economy could look like, complete with detailed and useful calculations on what's involved," says Catherine Wolfram, associate professor at University of California (UC) at Berkeley's Haas School of Business and co-executive director at the university's Center for Energy and Environmental Innovation.

Clean Energy 2030 is seeking to achieve these and other goals by suggesting specific actions in three key areas: energy efficiency, renewable electricity, and transportation.

Energy Efficiency

The plan addresses the demand side of the energy equation by challenging the federal government and states to a "long-term commitment to energy efficiency" through such programs as national efficiency standards, building codes, "decoupling" of utility profits from sales, among others. The plan also proposes the large-scale deployment of a "smart" electricity grid. As a result, Google projects that energy demand can be kept flat at the 2008 levels (rather than the expected 1% annual demand increase). No doubt that Google is aware that California, its home state, over the last 35 years has decoupled from national trends of electricity demand as a result of landmark energy efficiency policies reducing its per capita requirements to 40% below the national average.

Electricity from Renewable Energy Sources

At the core of Google's renewable electricity proposal is the goal of completely eliminating coal and oil in electricity generation. While very little oil is used to make electricity, coal accounts for about half of all domestically generated electricity. In order to achieve this goal, Clean Energy 2030 calls for a substantial production increase of three key renewable energy sources: wind (onshore and offshore), solar (photovoltaic and concentrating solar power), and geothermal (conventional and enhanced).

Wind would provide the largest source of electricity with a total production of 380 GW by 2030 (up from 20 GW today). This would equate to wind serving nearly a third of total projected electricity demand in 2030. The plan projects the largest growth for solar power, which generates 1 GW today and would provide 250 GW by 2030 and serving about 12% of expected demand. Finally, geothermal power would grow from 2.4 GW production today to 80 GW in 2030.

Google projects modest expansions for other non-fossil energy sources, mainly nuclear (115 GW), hydro (75 GW), and biomass and municipal waste (23 GW), exactly as projected by the US Department of Energy's Energy Information Administration. The combination of these six energy sources would meet about 90% of expected consumer demand, with the remainder being serviced by natural gas (290 GW).

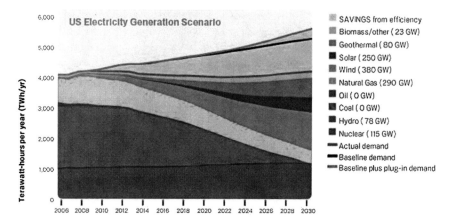

Fig. 9.3 Clean Energy 2030's proposed US Electricity Generation Scenario (*Source*: google.com/energyplan)

Transportation

The Clean Energy 2030 plan's initiative on transportation focuses solely on personal vehicles, which today account for about 60% of the sector's fuel consumption and CO_2 emissions. The plan seeks to address transportation with two key strategies. First, it calls for a significant increase in plug-in vehicle sales, with expected annual sales at 16.5 million cars by 2030. Second, the plan assumes that conventional vehicles will reach efficiencies of 45 mpg (the average fuel efficiency target mandated in Europe by 2012).

Economics and Jobs

Clean Energy 2030 also includes an outline of the costs as well as the savings and jobs that would be generated by all the proposed strategies of the plan. The plan estimates a discounted net savings of $211 billion (or $820 billion undiscounted). While calculating the economic impact of such a complex and long-term plan is difficult, Google has clearly outlined the assumptions they have made to arrive at that number.

The plan also predicts a positive net impact on 9 million jobs. This number is based on a number of different studies and includes "direct jobs (e.g., construction and operations of the power plants) as well as 'indirect' jobs in associated industries (e.g., accountants, lawyers, steel workers, and electrical manufacturing) and 'induced' jobs through economic expansion based on local spending." The authors consider this estimate to be a conservative number, because it accounts for growing efficiencies realized in the sector that would result in a trailing-off of the job growth rate over time.

The calculations expect the greatest proportion of jobs to continue to be in efficiency and wind, while the solar (both PV and CSP) and geothermal industries will also see dramatic increases.

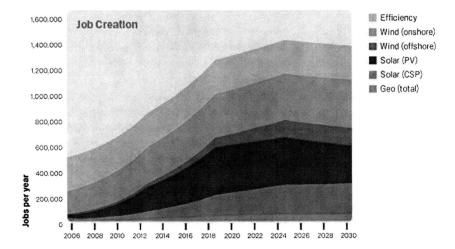

Fig. 9.4   Job creation (*Source*: Google.com/energyplan)

A Time for Action

In its introduction, the Clean Energy 2030 plan recognizes a variety of other plans, including the plan presented by President Barack Obama during his campaign, the Natural Resources Defense Council, McKinsey & Company, and former Vice President Al Gore's. The number of proposals as well as the variety of solutions they present speak to the fact that at this point in time no person or entity knows what the right path will be. Adds Professor Wolfram, "No one should mistake the plan for a solution, as it does not spell out how we get from here to there. Developing policies to set up the correct incentives is an important but tricky part of the process."

These plans as well as the organizations advocating energy and climate bills in the U.S. Congress make the case that now is the time to take action. As Dan Reicher, Google's director of climate and energy initiatives and a former assistant secretary of energy, puts it, "Solving both our economic and energy crises will be at the top of the list for the new Administration and Congress. We have an unprecedented opportunity to simultaneously create millions of high-paying green jobs while developing renewable energy and reinventing our electric grid. We are working with stakeholders in Washington and beyond to advance key policies in support of Clean Energy 2030 goals and are thrilled to see policymakers debate ideas from smart grid and plug-in hybrid vehicles to renewable energy tax credits and efficiency standards. The moment for action cannot come soon enough."

Sources

- Google Clean Energy 2030 Plan: http://www.google.com/energyplan.
- American Wind Energy Association, AWEA 2nd Quarter 2008 Market Report, 2008: http://awea.org/publications/reports/2Q08.pdf.
- Clean Edge, Utility Solar Assessment Study: Reaching Ten Percent Solar by 2025, 2008: http://www.cleanedge.com/reports/reports-solarUSA2008.php.
- Energy Information Administration, Table 8.11a, Electric Net Summer Capacity: Total (All Sectors), Selected Years, 1949–2007, Annual Energy Review, US Department of Energy, 2007: http://www.eia.doe.gov/emeu/aer/pdf/pages/sec8_42.pdf.
- Energy Information Administration, Table 9.2, Nuclear Power Plant Operations, 1957–2007, Annual Energy Review, US Department of Energy, 2007: http://www.eia.doe.gov/emeu/aer/txt/stb0902.xls.
- Energy Information Administration, Existing Electric Generating Units in the United States, 2005, US Department of Energy, 2007. http://www.eia.doe.gov/cneaf/electricity/page/capacity/existingunits2005.xls.
- Geothermal Energy Association, All About Geothermal Energy: Employment: http://www.geo-energy.org/aboutGE/employment.asp.
- Interlaboratory Working Group, Scenarios for a Clean Energy Future, Oak Ridge National Laboratory and Lawrence Berkeley National Laboratory, ORNL/CON-476 and LBNL-44029, 2000: http://www.ornl.gov/sci/eere/cef/.
- Krupp, F. and M. Horn, Earth: The Sequel. The Race to Reinvent Energy and Stop Global Warming, New York: Norton, 2008: http://earththesequel.edf.org/.
- Laxson, A., M.M. Hand, and N. Blair., High Wind Penetration Impact on U.S. Wind Manufacturing Capacity and Critical Resources, National Renewable Energy Laboratory, NREL/TP-500-40482, 2006: http://www.nrel.gov/docs/fy07osti/40482.pdf.
- Massachusetts Institute of Technology, The Future of Geothermal Energy: Impact of Enhanced Geothermal Systems (EGS) on the United States in the 21st Century, DOE Contract DOE-AC07-05ID14517, 2007: http://geothermal.inel.gov/publications/future_of_geothermal_energy.pdf.
- Nadel, S., Energy Efficiency and Resource Standards: Experience and Recommendations, American Council for an Energy-Efficient Economy, Report E063, 2006: http://www.aceee.org/pubs/e063.htm.

- Navigant Consulting, Economic Impacts of Extending Federal Solar Tax Credits, Final Report Prepared for the Solar Energy Research and Education Foundation, 2008: http://www.seia.org/galleries/pdf/Navigant%20Consulting%20Report%209.15.08.pdf.
- National Renewable Energy Laboratory, Job and Economic Development Impact Models, 2008: http://www.nrel.gov/analysis/jedi/.
- Prindle, B., Eldridge, M., Laitner, J. A., Elliott, R. N., and S. Nadel, Assessment of the House Renewable Electricity Standard and Expanded Clean Energy Scenarios, American Council for an Energy-Efficient Economy, Report E079, 2007: http://www.aceee.org/pubs/e079.htm.
- Simons, G. and J. McCabe, California Solar Resources in Support of the 2005 Integrated Energy Policy Report, Draft Staff Paper, CEC-500-2005-072-D, 2005: http://www.energy.ca.gov/2005publications/CEC-500-2005-072/CEC-500-2005-072-D.PDF.
- US Department of Energy, 20% Wind Energy by 2030: Increasing Wind Energy's Contribution to U.S. Electricity Supply, DOE/GO-102008-2567, 2008: http://www.20percentwind.org/20percent_wind_energy_report_05-11-08_wk.pdf.
- Blinder, Alan S., A Modest Proposal: Eco-Friendly Stimulus, The New York Times, July 27, 2008: http://www.nytimes.com/2008/07/27/business/27view.html.
- Godoy, M. CAFE standards: Gas-Sipping Etiquette for Cars, National Public Radio, 2007: http://www.npr.org/templates/story/story.php?storyId=5448289.
- Neff, J., Lutz says new CAFE standards will increase car price by $6 k, Auto Blog Green, 2008: http://www.autobloggreen.com/2008/01/13/lutz-says-new-cafe-standards-will-increase-car-price-by-6 k/.
- Union of Concerned Scientists, Fuel economy basics: http://www.ucsusa.org/clean_vehicles/solutions/cleaner_cars_pickups_and_suvs/fuel-economy-basics.html.
- Electric Power Research Institute, The Power to Reduce CO_2 Emissions, 2007: http://mydocs.epri.com/docs/public/DiscussionPaper2007.pdf
- Energy Information Administration, Annual Energy Outlook, US Department of Energy, 2008: http://www.eia.doe.gov/oiaf/aeo/.
- Gore, Al, The Climate for Change (op-ed), The New York Times, November 9, 2008: http://www.nytimes.com/2008/11/09/opinion/09gore.html.
- Intergovernmental Panel on Climate Change, Fourth Assessment Report, 2007: http://www.ipcc.ch.
- Intergovernmental Panel on Climate Change, IPCC Special Report on Carbon Dioxide Capture and Storage. Prepared by Working Group III of the Intergovernmental Panel on Climate Change [Metz, B., O. Davidson, H. C. de Coninck, M. Loos, and L. A. Meyer (eds.)]. Cambridge University Press, Cambridge, United Kingdom and New York, NY, USA, 442pp., 2005: http://www.ipcc.ch/ipccreports/srccs.htm.
- Marland, G., T. Boden, and R. J. Andres, Global, Regional, and National Annual CO_2 Emissions from Fossil-Fuel Burning, Cement Production,

and Gas Flaring: 1751–2005, Carbon Dioxide Information Analysis Center Environmental Sciences Division, Oak Ridge National Laboratory, 2008: http://cdiac.ornl.gov/ftp/ndp030/global.1751_2005.ems.

- McKinsey & Company, Reducing US Greenhouse Gas Emissions: How Much at What Cost? 2007: http://www.mckinsey.com/clientservice/ccsi/pdf/US_ghg_final_report.pdf.
- Natural Resources Defense Council, A Responsible Energy Plan for America, 2005: http://www.nrdc.org/air/energy/rep/rep.pdf.
- Obama for America web site, New Energy for America, 2008: http://my.barackobama.com/page/content/newenergy.
- Pacala, S. and R. Socolow, Stabilization Wedges: Solving the Climate Problem for the Next 50 Years with Current Technologies, Science, 305, 968, 2004: http://www.princeton.edu/wedges/articles/ (other "wedges" articles also available via this link).
- Pickens Plan: http://www.pickensplan.com/.
- WeCanSolveIt.org.

Chapter 10
Sustainable Communities in the UK

Claire Bonham-Carter

C. Bonham-Carter (✉)
EDAM AECOM, San Francisco, California, USA
e-mail: claire.bonham-carter@aecom.com

An examination of how government policy has enabled sustainable best practices to be achieved in recent, current, and future community development projects, including Upton, Bickershaw, and Peterborough.

Requirements of Sustainable Communities

Integration of Best Practices

In the pursuit of sustainability, we must rethink built environments according to their impacts on natural and social environments. Major aims in the present global circumstances include both mitigating and adapting to climate change, ensuring the security of long-term water and energy resources, protecting and restoring the planet's biodiversity, cleansing ecosystems damaged since the industrial age, providing for economic growth and security, fostering social equity and cohesion, and preserving and enhancing cultural diversity. All of these things must be taken into consideration holistically, and ultimately, we must provide appealing, safe, and comfortable places for people to live.

A fundamental shift in development patterns is necessary to achieve these aims. The principal components of sustainable communities are as follows:

- Density and transit – development must be concentrated around mass transit routes to create communities in which low-carbon transportation is the most attractive option, open space is conserved, social cohesion is promoted, and energy sources can be shared to the greatest possible extent.
- Mix of uses – incorporating retail, residential, recreation, and social infrastructure allows people to spend most of their time within their local neighborhood, reducing the need for transit and improving a sense of community.
- Variety of housing types and tenures – in addition to architectural variety, this allows for a community of multiple ages and socioeconomic strata.
- Walkability and cyclability – ensuring safe and attractive routes to places people want to go, which decreases the need for automobiles, ties a community together, and promotes a healthy and happy lifestyle.
- Water-sensitive urban design – this practice embraces the entire urban water cycle, including stormwater management, wastewater minimization, and potable water conservation.
- Energy efficiency – reducing energy demand as much as possible through appropriate design, and supplying remaining needs as efficiently possible through low- or zero-carbon techniques with a focus on providing community-level systems wherever possible.
- Ecology and open space – the provision of open space enhances the built environment, reduces the heat island effect, enables water-sensitive urban design, provides the potential for carbon sequestration, and preserves habitat primarily for wildlife but also for recreation.
- Public realm – the community must have shared iconic spaces that draw people together and establish a memorable and distinctive sense of place.
- Cultural responsiveness – communities must be planned and designed in dialogue with their adjacent communities and their cultural and historic heritage, the modern building drawing from rather than replacing the past.

- Distinctness of character – in achieving these aims, communities must each be their own place, recognizable and distinct from others.
- Community governance – the community has an important role in contributing to long-term management and governance of the place.

Ultimately, sustainability depends on the integration of all of these elements to achieve an organic system. It is about functional cooperation and mutual benefit. It is something that has been achieved in small pieces and with varying levels of success. We are engaged in a process of learning, collaborating, and continually improving.

Public Policy

The sustainability principles outlined above provide the framework for an ideal new community. These are not new and have been practiced either individually or as a package on exemplar projects in many parts of the world. However, implementing this framework as a standard on all mainstream developments is not feasible without the mandate of public policy and legislation.

In the United Kingdom, an entire suite of planning policy and building regulation changes have been critical in motivating the development industry's response to climate change. This process could be said to have started formally with the publication of the government's vision for sustainable communities, set out in *Sustainable Communities – Building for the Future* in 2003. This was closely followed by the publication of the *Planning Policy Statement 1 on Delivering Sustainable Development* published in 2005 (which replaced *Planning Policy Guidance 1: General Policies and Principles*), which acknowledged that sustainability should be the bedrock for all new development and drew together principles from other existing pieces of guidance.

This document opens by stating that "Sustainable development is the core principle underpinning planning." The four key elements of the statement are as follows:

- Social cohesion and inclusion
- Protection and enhancement of the environment
- Prudent use of natural resources
- Sustainable economic development

The statement goes on to explain that "These aims should be pursued in an integrated way through a sustainable, innovative and productive economy that delivers high levels of employment, and a just society that promotes social inclusion, sustainable communities and personal well being, in ways that protect and enhance the physical environment and optimize resource and energy use."

The government states that regional planning bodies and local planning authorities should ensure that development plans contribute to global sustainability by addressing the causes and potential impacts of climate change – through policies that reduce energy use, reduce emissions (e.g., by encouraging patterns of development that reduce the need to travel by private car, or reduce the impact of moving freight), promote the development of renewable energy resources, and take climate change impacts into account in the location and design of development. There is language within the statement, which gives regional planning authorities and local authorities the power to promote resource and energy-efficient buildings; community heating schemes, the use of combined heat and power, and small-scale renewable and low-carbon energy schemes in developments; the sustainable use of water resources; and the use of sustainable drainage systems in the management of run-off.

This policy statement is supplemented by a range of more detailed guidance on transportation, biodiversity, renewable energy, sustainable waste management, and development and flood risk. However, to further strengthen the overarching statement – and recognizing the very real threat of climate change, which the government believes is the greatest long-term challenge facing the nation today – a *Supplement to Planning Policy Statement 1: Planning and Climate Change* was published in December 2007. This sets out how the government expects planning authorities to ensure that new developments contribute to reducing emissions and to stabilizing climate change while taking into account the already unavoidable consequences.

Of particular note are the sections that ask for subregional and local planning to focus substantial new development on locations with good accessibility by means other than the private car and where energy can be gained from decentralized energy supply systems or where there is clear potential for this to be realized in the future. This mandate recognizes the significant proportion of carbon emissions that comes from single occupancy automobile use to, from, and within new communities and the key role that new and improved public transportation must play in any new development.

Energy supply to new communities is also critical, and this document gives authorities the power to request a proportion of energy to be supplied from decentralized and renewable or low-carbon energy sources. This decision follows the examples of local authorities across the UK requiring a minimum of 10% of energy for a new development to come from renewable sources, a proportion that has risen to 20% in some places, London among them. Permits cannot be granted unless this provision is met. The focus on decentralized energy sources recognizes the inefficiency of grid-generated electricity, with additional losses caused by transmission lines stretching hundreds of miles in length. Locally generated power can be cheaper, lowers carbon, helps to balance peak loads, and provides fuel flexibility.

There are examples of local government requiring major developments to consider heating, cooling, and powering through localized sources such as a district energy center fueled by a combined heat and power (CHP) plant. London, for example, aims to supply 25% of its energy through decentralized means by 2025, connecting both existing and new development to highly efficient heating and power networks. Major mixed-use development is unlikely to receive permits unless such a network is a key piece of the development plan, or unless it can be demonstrated by the applicant, with regard to the type of development involved and its design, that this is not feasible or viable. A study carried out on behalf of the Greater London Authority showed that average developments submitting applications since the adoption of the low- and zero-carbon policies had been designed to achieve a reduction in carbon emissions of up to 25% below current Building Regulations.

Supplement to Planning Policy Statement 1: Planning and Climate Change

Local requirements for decentralized energy to supply new development
Paragraph 27: In considering a development area or site-specific target, planning authorities should pay particular attention to opportunities for utilizing existing decentralized and renewable or low-carbon energy supply systems and to fostering the development of new opportunities to supply proposed and existing development. Such opportunities could include co-locating potential heat customers and heat suppliers. Where there are existing decentralized energy supply systems, or firm proposals, planning authorities can expect proposed development to connect to an identified system or be designed to be able to connect in future. In such instances and in allocating land for development, planning authorities can set out how the proposed development would be expected to contribute to securing the decentralized energy supply system from which it would benefit.

Carbon emissions are not the only focus of the Planning and Climate Change Supplement. The urban heat island effect is here referenced for the first time in planning policy. The statement speaks to the contribution that can be made by open space and green infrastructure to provide urban cooling as well as sustainable drainage systems and to conserve and enhance biodiversity (recognizing the need to adapt to likely changes in climate).

The Building Regulations are another important legislative driver, and the document *Building a Greener Future* sets out a progressive tightening of Building Regulations to require major reductions in carbon emissions from new homes to reach zero carbon by 2016. By 2010, standards will be set to achieve a reduction in carbon emissions of 25%; in 2013 standards will be set to achieve a reduction of 44%. By 2016, all new homes will be zero carbon (by 2019 for nonresidential buildings). A third but closely linked legislative driver was issued at the same time as the *Building a Greener Future* consultation – the Code for Sustainable Homes – which has replaced Ecohomes as the assessment methodology for new homes in the UK. The energy credit section of the code follows the same carbon emission reduction progression as planned for the Building Regulations, laying out a clear route map for the development industry and an opportunity to be rewarded if they achieve the targets in advance. See box for more details.

Code for Sustainable Homes

In April 2007 the Code for Sustainable Homes replaced Ecohomes for the assessment of new housing in England. The code is an environmental assessment method for new homes based on <u>Ecohomes</u> and contains mandatory performance levels in seven key areas:

- Energy efficiency/CO_2
- Water efficiency
- Surface water management
- Site waste management
- Household waste management
- Use of materials
- Lifetime homes (applies to code level 6 only)

The code has a scoring system of six levels. The different levels are made up by achieving both the appropriate mandatory minimum standards together with a proportion of the "flexible" standards. Code assessments are carried out in two phases:

- An initial assessment and interim certification is carried out at the design stage
- Final assessment and certification are carried out after construction

On February 27, 2008, the government confirmed that from May 1, 2008, it would be mandatory for all new homes to have a rating against the code.

These documents establish a long-term policy and regulatory framework that will give industry the certainty they asked for and the time to adjust, as well as the incentive to innovate and invest in the technologies needed, driving down costs over time and enhancing competitiveness. It is expected that the higher standards and new designs and technologies will bring benefits in reduced fuel bills as well as reductions in carbon emissions. The government recognizes that this will involve increased upfront costs for these new homes, although it is expected that these will be reduced over time as new technologies develop and economies of scale are achieved.

Education and Marketing

Marketing takes on a unique importance with sustainable communities. The traditional housing development seeks to market itself only as an appealing place to live, with a focus on the individual dwelling. For a sustainable community, however, the focus broadens out to encompass the benefits of the mix of the neighborhood as well as the other more traditional environmental elements of sustainability providing appeal. Sustainability thus becomes a selling point but only if it can be readily explained. The transparency of sustainable planning and design is essential in winning the demand of potential residents. The more widely a community can be marketed, in turn, the better it serves as a regional, national, or global exemplar, in increasing market demand for a sustainable lifestyle, in demonstrating the viability of such a program to the development industry, and in providing inspiration and competition for planners and designers. In the case studies below, we will see several forms of dynamic marketing.

Case Studies in the UK

Upton

The UK established English Partnerships (EP) as its national development agency in May 1999 by combining the Commission for New Towns (established 1961) and the Urban Regeneration Agency (established 1993). EP states their aim as "to achieve high-quality, well-designed, sustainable places for people to live, work and enjoy".[1]

Specific objectives include increasing the amount of high-quality affordable housing available, making the best use of the UK's scarce supply of land, reducing low-demand and abandoned housing, delivering and increasing skills in urban regeneration, increasing private-sector investment in housing, and promoting best practices in both urban design and construction methods. EP is empowered to acquire and assemble land in areas targeted for development and regeneration and to guide those processes sustainably in conjunction with local authorities.

The spatial strategy for the East Midlands identified Northampton as the major population and employment center in the south of the region. In 1997, prior to the establishment of EP, the Commission for New Towns applied for an Outline Planning Approval for Upton, an extension of Northampton on its southwest fringe. The approval was granted in May 2000, which triggered the partnership arrangement

---

[1] EP have now changed their name to the Housing and Communities Agency and have a new web site, although the old one is still there: www.homesandcommunities.co.uk (http://www.englishpartnerships.co.uk/about.htm).

between EP as land owners, Northampton Borough Council, and The Prince's Foundation to update the plan and deliver Upton. This update was the result of an Enquiry by Design public consultation held in December 2001 that established high objectives for Upton as a sustainable community.

Upton is a 44-ha greenfield site bounded on the north and east by roads and on the west by the Upton Park expansion area. To the south of Upton is the Upton Country Park, which includes the flood plain of the River Nene. Aiming to deliver 1,382 homes by its completion in 2013, Upton was conceived as a mixed-use development offering the full range of dwelling types supported by economic opportunity and social infrastructure. With the commercial area located along the frontage to the north, the development density was envisioned to transition gradually toward the countryside to the south and west. Environmental sensitivity and energy efficiency were established early on as key considerations.

Engaging the local community was a fundamental aspect of the project. The working group consists of the partners – EP, Northampton Borough Council, and The Prince's Foundation – and the consultants – EDAW (master planners, landscape designers), Alan Baxter Associates (transport engineers), Quartet (landscape implementers), and Pell Frischmann (engineering implementers). Local residents and representatives from the Upton Parish Council joined the working group to form the Steering Committee. An ongoing dialogue among partners, consultants, and local residents thus informed the design of the development and ensured that the community would have ownership of their environment.

The 1997 Outline Planning Application resulted in a conventional vision for the new community. It would be a residential development, located around a community center and a school, car oriented, and would isolate the affordable housing in certain areas. The Urban Framework Plan that emerged from the 2001 Enquiry by Design process embraced a more innovative, sustainable approach. The number of residential units was increased 35% to create a critical mass of local population that will help support local facilities. The retail uses were moved to the north of the development alongside main route into Northampton to ensure commercial viability. Pedestrian-friendly street design was incorporated, as was a sustainable approach to stormwater management. The affordable housing was also to be integrated rather than segregated. Overall, the plan aspired to a socially cohesive, environmentally sustainable community.

To implement this aspiration, design codes were used as an effective tool. These established a hierarchy of connectivity, from primary arteries to streets with stormwater management features to other secondary streets, lanes, and mews. While cars have been accommodated, almost everything in the development is walkable, and the size and openness of the streets favors pedestrians as well promoting safety by offering maximum visibility in many directions.

The codes also establish a hierarchy of design types. Addressing aspects such as scale of buildings, architectural detail, fenestration, building types, and materials, the codes identified a gradient of different character areas. The Urban Boulevard along Weedon Road would have a high density (60 dph), consisting of a mixed-use commercial center with dwellings above the street. The Neighborhood Spine, reaching southward through the center of the development, would have a reasonably high density (45–50 dph), with space for townhouses, public transport, the school, and open spaces. The Neighborhood Edge, on the southwest border of the development, would have low density (30 dph), with rural dwellings situated on informal lanes.

The Neighborhood General, embracing the rest of the allotted land, would have a medium density (35–40 dph), consisting of residential communities on streets and mews.

High energy and water efficiency of the houses was a major aim; all were to achieve BREEAM Excellent rating. All homes achieve the same level of sustainability in an overall sense, but they do so through different combinations of technologies according to the various developers' responses to site-specific development briefs. Some developers are particularly keen, and they teamed up with architects with strong credentials in zero-carbon homes in the UK. As a result, there are a small number of homes in Upton achieving level 6 for the Code for Sustainable Homes (see box later for explanation).

Fig. 10.1 Upton's streets are spacious and pedestrian oriented. Its design is consistent and responsive to context while incorporating variation

BREEAM (Building Research Establishment Environmental Assessment Methodology)

BREEAM was the first environmental certification scheme for buildings, established in 1990 in the UK, initially just for offices but now with specific schemes for retail, industrial buildings, education, prisons, healthcare, homes (originally "Ecohomes" now replaced by the Code for Sustainable Homes), multi-residential, and a bespoke scheme to cover other building types such as laboratories, hotels, and halls of residence. The scheme covers eight sustainability issues including energy, water, materials, and waste, transport, health and well-being, management, pollution, and land use and ecology. Issues are weighted according to their relative environmental impact. The building is then rated on a scale of Pass, Good, Very Good, and Excellent. The United States Green Building Council Leadership in Energy + Environmental Design (USGBCLEED) scheme was based on BREEAM.

Roughly 85% of all homes employ new energy and water technologies. Although the remaining 15% have achieved BREEAM Excellent, they were not asked to include any such technologies. However, one of the many ways the environmental targets are being achieved is through the use of Redland PV 80 photovoltaic tiles from Lafarge Roofing that will provide southerly facing homes with around 960 kWh/year – providing a third of the typical family's electricity demand (about 20% of a typical Californian electricity demand) and reducing carbon dioxide emissions by 400 kg/year. The orientation of buildings to take advantage of sunlight maximizes the effectiveness of solar energy. It is hoped that the pedestrian emphasis of the development will further reduce Upton's overall carbon footprint by reducing the number of daily short car trips. All homes are designed to harvest rainwater, and all southerly facing homes have been equipped with solar water-heating systems. Some residences will feature green roofs; others will have heat recovery ventilation. Also featured are ground source heat pumps, communal biomass boilers, and low-flow faucets. These measures have exceeded the scores necessary to achieve the BREEAM Excellent rating.

Integrated stormwater management was another objective that emerged from the Enquiry by Design. Overlaid with Upton's street network is a sustainable urban drainage system (SUDS), or what in the US would be called low-impact development (LID) or best management practices (BMP). This consists of vegetation-filled swales that will collect, convey, and naturally process stormwater run-off, alleviating the pipe system of strain due to excess volume while relieving the watershed of contaminants, which are trapped in the soil beds of the swales. Upton sits on the edge of a flood plain, and SUDS is needed to alleviate the burden of the pipe system especially during short, sharp, and intense downpours, which are increasingly common in the UK. This urban hydrological system discharges water cleansed of pollution into the River Nene to the south, integrating Upton into its local ecology.

Upton's context, with parkland to the south and west, affords residents with the recreational and aesthetic benefits of the Upton Country Park. Recognizing these benefits, the master plan left these lands undeveloped and allowed green space to permeate the urban boundary, with much of the site remaining mostly rural and development clustered densely in specified areas. The Upton Country Park is about 4 ha. In addition to that, Ashby Wood, an existing copse, was integrated with the green SUDS ecological network, which creates an interconnected foraging network for rabbits and other small animals. A small barn has also been restored and relocated to the park to create a habitat for bats. Thus a balance is achieved wherein density fosters greater sense of community, makes urban functions more efficient, and allows green space to exist within and adjacent to the urban boundary.

Fig. 10.2 Solar panels on houses at Upton. Photovoltaic panels (electricity producing) and Solar thermal panels (hot water producing) on houses at Upton

Fig. 10.3 Upton's sustainable urban drainage system is integrated with its public spaces

Another major aim of the Urban Framework Plan was to deviate from the "clone-town" phenomenon that had been occurring in the UK. Variety was essential to the planning and design of Upton at all levels. Different housing types and tenures were mixed to promote cohesion among a community of diverse residents. The different characters represented throughout the development allow residents choices with regard to density and traditional versus contemporary design. For development purposes, the site was split into seven sections; a two-stage process selected the most suitable developers for each section, incorporating a range of creative styles. Upton's appeal as a place to live, work, and play directly relates to its sustainability. Sustainability is an attractive asset for some residents; others who are drawn to Upton for its character and amenities will come to embrace a sustainable lifestyle.

Upton was marketed through the creation of a video that explained its planning concepts and interviewed residents who expressed their satisfaction that these concepts had been realized. Using attractive graphics, pleasant music, and an iconic style, the video conveys a complex development program in simple manner.

Upton is scheduled to complete in 2013, though current market downturn may postpone completion date. To date, 547 of the 1,382 dwellings are complete. Current residents are conscious of what their community offers them and what it offers the world as an exemplar. They feel safe in an open environment with wide views. They enjoy walking and cycling to local amenities. They enjoy the parklands within easy access. They are proud of the solar panels displayed conspicuously on the fronts of homes rather than the rears. Stormwater moves safely and naturally through their domain. The development was the winner of the RTPI Sustainable Communities Award 2007.

The Carbon Challenge: A Technical Challenge

Building on the success of Upton and other projects, EP launched the Carbon Challenge in 2007 in an effort to accelerate the house-building industry's response to climate change. In this endeavor, EP sought to reach the highest level of the Code for Sustainable Homes without compromising design quality or cost efficiency. The aims of the program are to:

- deliver high-quality design combined with exceptional environmental performance;
- deliver innovative housing to level 6 of the Code for Sustainable Homes, BREEAM Excellent for commercial and community buildings, and CEEQUAL Excellent for civil engineering works;
- encourage residents to embrace a healthy and sustainable lifestyle that reflects the ethos of the scheme and advantage of its location;
- agree to share the lessons learnt with the wider house building and construction industry; and
- work on an open-book basis and commit to an evaluation and monitoring protocol.

EP developed a standard Carbon Challenge Brief, which is structured into five sections, covering delivering quality places, quality homes, quality building, Code for Sustainable Homes Code 6 (see below), and construction efficiency. In addition, each site has a site-specific brief covering planning and development terms to help developers achieve the high targets required for each particular site.

The Code for Sustainable Homes addresses the sustainability of a residential unit as a complete package, examining a wide range of aspects and strictly evaluating subjects according to six levels. The Carbon Challenge has mandated that homes achieve the sixth and most difficult level, which from an energy perspective requires a home to be net zero carbon – which means that the energy generated from fossil fuels must be less than or equal to that generated through renewable technologies. The UK government intends that all new homes will be net zero carbon by 2016 (compared to the state of California, which is aiming to achieve the same target by 2020, with commercial development achieving the same target by 2030). Stringent targets relating to water and material use are also included.

Carbon Challenge Sites: Bickershaw and Peterborough

The development of Bickershaw, near Wigan, not only provided one of four Carbon Challenge sites, but also is an EP National Coalfields Programme site. Brownfield development raised the inherent sustainability of the project, as land previously degraded during the industrial age was to become a high-quality, sustainable, residential community with a commercial core set around a marina. The 18.3-ha site is to deliver leisure, cafes and restaurants, shops, offices, and also around 650 homes to code level 6 as well as other Carbon Challenge standards.

Following the success of the Upton Design Codes, the design code method was employed at Bickershaw as well, where it would be even more necessary in achieving the strict energy-efficiency standards without sacrificing other community objectives. The codes formed an effective tool for translating sustainability into design quality and then communicating this translation to both developers and the community so that all could collaborate on a mutually understood vision.

Designers took an elemental approach to Bickershaw, identifying earth (materials and ecosystems), air (wind and ventilation), community (health and lifestyle), water (supply and drainage), and fire (energy and sunlight) as the fundamental components. It was crucial to consider the interrelationships among these elements rather than any one in isolation. Thus the codes reflect Bickershaw as a cohesive web, an organic system. For Bickershaw, a web site is being created as a marketing and education tool, specifically targeted at developers.

The second Carbon Challenge site is located on the south bank of the city of Peterborough. Here on a 7-ha parcel of land, 344 new homes will be delivered, 35% of them affordable housing. EP, East of England Development Agency, and Peterborough City Council have teamed up for the project, which will include creative SUDS features that give character to public spaces, a "green spine" with orchards and allotments, canopy walkways, and 650 m^2 of retail floor space. The community will be encouraged to grow its own food and to buy local food products.

The development is intended to fit into four contexts: its local fabric, providing a mix of uses that fulfill community needs; its urban fabric, helping Peterborough to be the UKs "Environment Capital" for sustainable living and technological excellence; its national fabric, showing developers that zero carbon and good design can both be met; and its global fabric, as an exemplar in combating climate change.

The site is bounded by a railroad track and a river to the north, a Horse Fair car park and main road to the east, a parkway to the west, and a Football Club to the south. The Carbon Challenge development is intended to be a bridge or a point of connection in more ways than one. A sizable city that has been growing ever since the advent of the railroad, Peterborough's urban core now extends to the north bank of the river. The development on the south bank must allow for the expansion of the core across the river, incorporating dense commercial areas into its typologies. Extensive green space exists east and west of the city; the South Bank development provides the opportunity to link these areas both recreationally and ecologically through pedestrian and cycle connectivity and the use of measures such as SUDS, wetlands, green roofs and walls, and nesting boxes.

Transit is a key consideration in the South Bank development, which will establish the patterns by which Peterborough will expand into the future. Pedestrian and cycle connectivity as well as light rail must be enabled if the car is to be outmoded. One challenge of the site is a heavy rail line dividing the area from the river, with crossings only at the two roads at either end of the site. For the city core to extend across the river in a sustainable fashion, the engineering challenge of bridging the rail line for pedestrians and cyclists at suitable sites, which are limited, must be met.

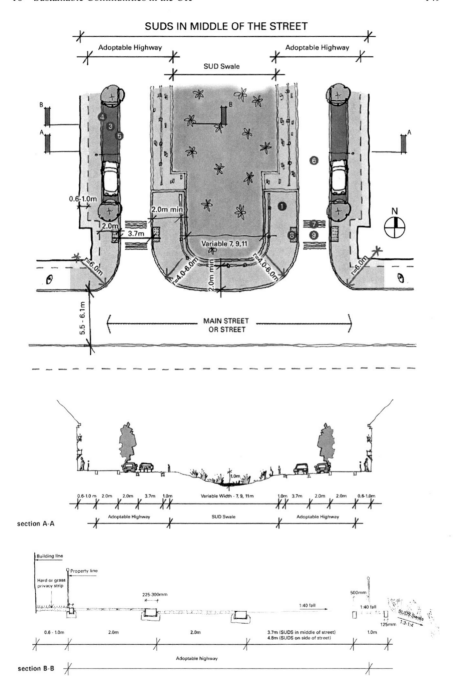

Fig. 10.4 The Bickershaw Design Codes aid developers in implementing sustainable best practices – detail on streetscape

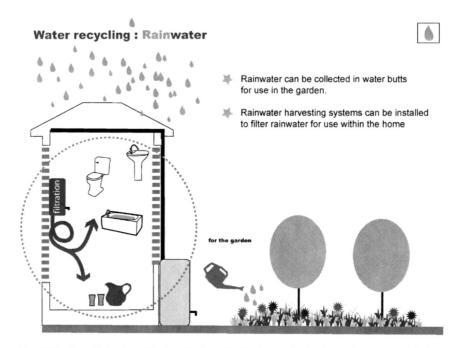

Fig. 10.5 The Bickershaw Design Codes aid developers in implementing sustainable best practices – detail on rainwater harvesting

In the effort to promote long-term economic sustainability, the Carbon Challenge calls for homes to be constructed so that they may in the future be easily converted to retail space as necessary. Design measures that ease this transition include higher ground floors, flexible frontages that can be dismantled without weakening structure, more open ground-floor plans, and larger windows. The South Bank development will employ these measures in the denser of its two character areas. The other area will be more suburban in form, with semi-detached, terraced dwellings, or town houses. Space would be more abundant here, encouraging the use of gardens. A gradient of building heights will also be used, with six or more stories in gateway areas, such as the area fronting the London Road and the central area where the pedestrian/cycle bridge over the railway may be located. The rest of the urban area would feature six-story buildings, with the suburban area featuring four-story buildings.

The grouping and orientation of buildings must balance a variety of objectives, including the establishment of landmarks and views that promote community and enable way finding, the maximum effectiveness of solar energy, the creation of common outdoor spaces, the potential for using a district heating (and power) system, and the energy efficiency of buildings. The greater the number of shared walls, the less the heat lost to the outdoors. Similarly, the design of streets, which must be considered as places, as opposed to roads, which merely conduct vehicular traffic, must accommodate purposes including public realm, stormwater management, shared

Fig. 10.6 Bickershaw incorporates a network of green space and integrated stormwater management

utilities, green space, and habitat. A corridor width of 25–30 m has been planned to accommodate these uses. A hierarchy of Main Boulevard, to Main Avenue, to Lanes, to Pedestrian and Cycle Ways has also been established. Overlaid in this hierarchy is one of green space, incorporating recreational uses, habitat, and stormwater management.

A net zero carbon strategy will need to be developed for the site, and to assist in this process, a thorough analysis of the potential for a range of zero- and low-carbon technologies has been undertaken covering combined heat and power, both gas and biomass fired, district heating systems, wind turbines, and solar power. The opportunities for some renewable technologies are possibly constrained due to the site size and location. For example, the site itself is unlikely to be suitable for a large-scale turbine due to the density of development required, but the adjacent open spaces to the north and east of the site warrant further investigation. Developers will need to think at an early design stage about the location of an energy center to supply a district heating and power network in terms of access, particularly if that energy center could be fueled by local biomass.

A wet retention pond will be placed in one of two sites within the South Bank development, serving as the destination of the SUDS features. The features should be considered conspicuous landscape elements used to enhance public spaces through creative design and the opportunity to introduce vegetation. Rainwater harvesting should be incorporated as well with underground tanks for both public and private use. Stormwater should be considered a highly valuable asset, rather than a hindrance to be disposed of. The green links that house the SUDS features should interact with streets and houses, and pedestrian connectivity should intertwine with them while keeping public space viewable from houses to promote safety.

At present the consortium has selected through competition its preferred developer to bring the site forward, based on a detailed response to the design brief held in 2008. As of December 2008 they were working with the developers who are hoping to submit a planning application in 2009; unsurprisingly the market at time of writing is slowing things up a little.

Future Projects and Policy

Upton has become an exemplar for sustainable communities; Bickershaw and Peterborough, if built, promise to expand upon its model. Now an even higher level of sustainability is being pursued in the UK under the moniker of ecotowns.

In November 2008, a consultation on the draft Planning Policy Statement on ecotowns was launched. Ecotowns are new settlements that will have sustainability standards significantly above equivalent levels of development in existing towns and cities and that are separate and distinct but well linked to higher order centers and have sufficient critical mass to achieve the ecotown objectives. The idea behind the program is to try to help meet the challenge of climate change and housing growth. The program is designed to support a limited number of exemplar schemes to demonstrate what life could be like in a low-carbon future.

The government wants ecotowns to be exemplar projects that encourage and enable residents to live within environmental limits and in communities that are resilient to climate change. The design of ecotowns are being expected to take full account of their impact on local ecosystems, mitigating negative impacts as far as possible and maximizing opportunities to enhance their local environments. Ecotowns will be developments of a minimum of 5,000 homes – well linked to, but distinct from, existing settlements – that achieve the highest standards of environmental sustainability, including:

- employing renewable energy technologies
- exceptionally high quality of environmental building design and
- excellent public transport

The consultation document lays out specific requirements for issues including zero carbon, climate change adaptation, water, waste, green infrastructure, biodiversity, flood risk management, employment, and transportation. The ecotowns will be expected to exceed or fast-track building regulation standards.

The ecotowns initiative was developed with the aim of bringing forward up to 10 schemes and launching development by 2016. The government shortlisted 15 locations from 57 ecotowns bids in response to the ecotowns prospectus published in July 2007, from which 11 are still being considered for inclusion. This program is not without its detractors due to the greenfield and unconnected nature of some of the sites, but it has the potential to fast-track the widespread development of sustainable communities in the UK. This, in combination with the implementation of the Code for Sustainable Homes, the tightening of the Building Regulations, and the likelihood of carbon dioxide reduction goals being written into law in the UK within the year, means that widespread sustainable development has a strong chance of becoming reality in the UK within a decade.

Chapter 11
Sustainable Towns: The Case of Frederikshavn – 100% Renewable Energy

Henrik Lund and Poul Alberg Østergaard

H. Lund (✉)
Aalborg University, Denmark
e-mail: lund@plan.aau.dk

Henrik Lund is Professor in Energy Planning at the Department of Development and Planning at Aalborg University and Editor-in-Chief of the Elsevier international journal ENERGY. He holds a PhD in the implementation of sustainable energy systems (1990) and a senior doctoral degree in Choice Awareness and Renewable Energy Systems (2009). He was head of department from 1996 to 2002 and, for more than 25 years, he has researched and published books and papers about energy system analysis, energy planning, and energy economics. The International Energy Foundation (IEF) awarded him a gold medal for Best Research Paper within the area of Energy Policies & Economics in 1998. He has been involved in a number of research projects and committee works in Danish energy planning and in the implementation of various local energy projects.

Poul Alberg Østergaard is Associate Professor in Energy Planning at the Department of Development and Planning at Aalborg University. He teaches subjects ranging from the design of energy systems to planning methodologies and economics. His research focus is the integration of wind power in complex energy systems as well as the impacts on transmission grids. He is actively involved in the development of a number of energy plans for specific areas in Denmark. He holds the equivalent of a B.Sc. in electrical engineering, as well as a M.Sc.Eng. and a Ph.D. in energy planning (2000).

W.W. Clark II (ed.), *Sustainable Communities*, DOI 10.1007/978-1-4419-0219-1_11,   155
© Springer Science+Business Media, LLC 2010

Abstract A number of Danish energy experts in 2006 made the proposal that Denmark should convert a town to 100% renewable energy by 2015. The experts suggested Frederikshavn in the northern part of Denmark for a number of reasons: The town area of 25,000 inhabitants is well defined, the local support is high, and Frederikshavn already has several big wind turbines at the harbour. In February 2007 the city council unanimously decided to go for the project and set up a project organisation involving utilities and municipality administrators. Moreover, local industry and Aalborg University are involved in the project.

This chapter

- presents the methodology of mapping the existing energy system including transportation and defining the share of renewable energy, which is approximately 20% in the present situation.
- introduces a proposal for a potential 100% renewable energy system and a number of realistic short-term first steps by 2015, which will take Frederikshavn to approximately 40% by 2009 or 2010.
- describes detailed hour-by-hour energy system analyses of the proposal for a 100% RES system.
- relates the proposal to the perspective of converting Denmark to 100% renewable energy.

Introduction

This chapter provides an example of how Frederikshavn, within a short span of time, can be converted into a town with 100% renewable energy supply. The example is based on a proposal drawn up by a working group on a project called "Energy Camp 2006." The proposal is presented in detail in the paper "Next City, Frederikshavn – Denmark's Renewable Energy City."

It should be underlined that this is only a proposal, which may serve as an inspiration for future work. The final project must develop in close dialogue and cooperation with all the actors involved in converting this idea into reality. In addition, we must reassess the numerical basis of the study. This is especially necessary in the case of the energy demand of the industry and transportation sector, in which Danish mean figures are used. On the other hand, the calculations of electricity and heat demand are based on information provided by the supply companies of Frederikshavn.

Definition of Renewable Energy

Renewable energy is here defined as energy produced from natural resources such as sunlight, wind, rain, waves, tides, and geothermal heat, which are naturally replenished within a time span of a few years. Renewable energy includes the technologies which convert natural resources into useful energy services:

– Wind, wave, tidal, and hydropower (including Micro- and run-of-river hydro)
– Solar power (including photovoltaic), solar thermal, and geothermal
– Biomass and biofuel technologies (including biogas)
– Renewable fraction of waste (household and industrial waste)

Household and industrial wastes are composed of different types of waste. Some parts are regarded as renewable energy sources, such as potato peel, while other parts are nonrenewable energy sources, such as plastic products. Only the fraction of waste that is naturally replenished is usually included in the definition. In the EnergyTown Frederikshavn project, however, for practical reasons, the whole waste fraction is included as part of the renewable energy sources.

When calculating the share of renewable energy (RES) in the system, the import and export of power converts to the fuel needed in order to produce the power on a power plant with an efficiency of 40%. The same factor has been used when electricity production from wind power is compared to fuel. Moreover, when calculating the share of RES, wind power has been corrected to the expected production in a normal wind year. In Denmark the wind years vary within ±20%.

Development Phases

In Frederikshavn, the actual supply in 2007 has a renewable energy share of approximately 20%. Based on this fact, the project works with the following years and development phases:

1. The first step is until 2009 when the UN Climate Summit COP15 will be held in Denmark. The objective is to raise the share of renewable energy in Frederikshavn to approximately 40%.
2. Transformation to 100% renewable energy in Frederikshavn by 2015 on an annual basis. However, exchange of energy with the surrounding is allowed.
3. Further development of the 100% renewable energy system in such a way that possibilities are created for the transformation to 100% renewable energy in Denmark as a whole.

The distinction between phases 2 and 3 is caused by the fact that the purpose of the project is not to create an isolated "energy island" with no connections to the surrounding. On the contrary the purpose is to show how a 100% RES future will look like and what it will take to implement it. Consequently, for example, vehicles in Frederikshavn fueled by RES in 2015 may have to leave the town and may not be able to fuel in other parts of Denmark until the whole county is converted into 100 percent RES. And vehicles from outside Frederikshavn will still have to be able to fuel in the town. A sufficient quantity of "bio-gasoline" will be produced to cover the need for transportation in Frederikshavn, but not all cars will be expected to change to the use of bio-gasoline. Moreover, cars from Frederikshavn are not expected to be able to refuel with bio-gasoline at other locations in Denmark. Besides, exchange of electricity in and out of the town from one hour to another as well biogas by use of the natural gas network might occur. However, on an annual basis, the amount of fuels for vehicles, electricity and heat productions based on 100% renewable energy should meet the exact demands of Frederikshavn by 2015.

In 2030 the target is to implement a solution in which the amount of biomass resources and exchange of electricity and fuels will comply with a strategy in which the whole of Denmark is converted into 100% renewable energy. Again it is not a target to totally avoid exchange. However, the exchange should be limited to an appropriate level taking into consideration that some parts of Denmark may utilise more wind power than others, while other parts may utilise more biomass resources. Moreover, Scandinavia may explore mutual benefits of exchanging, for example, wind power in Denmark by hydropower in Norway.

The Present Situation: Year 2007, Approximately 20% Renewable Energy

The existing energy supply in Frederikshavn is shown in Fig. 11.1, including industry and transportation. The project covers Frederikshavn, a town of 25,000 inhabitants, and is equivalent to the existing local power supply. Most houses and apartments connect to public district heating, but the area also covers a small share of individually supplied homes.

The energy demand consists of:

– an electricity demand of 164 GWh/year supplied by the public grid
– a district heating demand of 190 GWh/year distributed into two seperate systems of 15 and 175 GWh respectively. Including net losses of 52 GWh/year, the annual production in 2007 added up to 242 GWh/year.
– a local transportation demand equal to 165 GWh/year of gasoline and diesel.
– heating of houses in individual boilers equal to 37 GWh of fuel and estimated 28 GWh of heat.
– 36 GWh fuel for industry, equal to an estimated heat demand of 28 GWh and process heat of 3 GWh.

Out of 7 GWh plus 6 GWh fuel for individual houses and industry, an estimated 70% can be converted into district heating.

The district heating in the large system (Frederikshavn) is produced partly on waste incineration (CHP) and partly on a combined heat and power (CHP) plant including peakload boilers fuelled by natural gas. The individual supply is based on oil- and gas-fired burners and a small amount of wood.

District heating in the small system (Strandby) is produced on a small CHP plant on natural gas.

Figure 11.1 shows that a large part of the power demand in Frederikshavn is met by the local wind power and CHP production. Additional to the electricity production on the three CHP plants, 10.6 MW of local wind power covers some of the demand. The rest is imported from the national grid. Here the latter is assumed to be produced on a coal-fired power station with an efficiency of 40% equal to the average of Danish power-only plants.

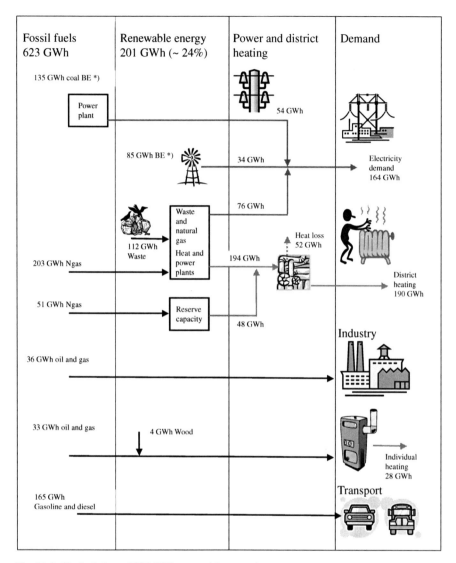

Fig. 11.1 Frederikshavn 2007 (20% renewable energy)
*Electricity as given in fuel equivalents of electicity production of a coal-fired steam turbine with an efficiency of 40%

The First Phase: Frederikshavn in the Year 2009

The year 2009 of phase one is chosen because the planned UN Climate Summit in Denmark will provide a good opportunity of promoting the project at an international level. The objective is to raise the share of renewable energy to approximately 40% by implementing the following four projects before the end of 2009:

– 12 MW wind turbines. The wind turbines are step one of a new offshore project of an expected 25 MW in total. The project has been decided, and the procedure of environmental impact assessments is in progress. The first 12 MW is expected to be implemented during 2009.
– 8000 m^2 of solar thermal in combination with additional 1500 m^3 of water heat storage and an absorption heat pump at the CHP plant at the small district heating supply of Strandby. At present the project is being implemented. The absorption heat pump will cool the exhaustion gas and raise the total efficiency from the present 94 to 98%, and the solar thermal will produce approximately 4 GWh of heat annually.
– Implementation of a facility to upgrade biogas from a local biogas plant outside the town to natural gas quality, and transport the gas into a biogas fuel station in Frederikshavn and invest in 60 bi-fuel cars. This will supply 7 GWh of biogas. The biogas that will not be used for transportation will be used in the CHP plant.
– Establishment of a 1 MW heat pump at the wastewater treatment plant of the town expected to utilise 2 GWh of electricity to take out 4 GWh of heat from the waste water and produce 6 GWh of heat for the district heating supply annually.

The total budget of this phase is approximately 200 million DKK (estimate), and as illustrated in Fig. 11.2, it is expected to raise the share of renewable energy to 38%.

The Second Phase Frederikshavn in the Year 2015: 100% RE on an annual basis

On a continuous basis, the EnergyTown Frederikshavn project is in the process of identifying a proper scenario for the implementation of a 100% renewable energy system by the end of year 2015. Here is a status on the results of such considerations. Each of the proposed projects will be subject to more detailed analyses in the coming period. However, the key components have been identified and are being planned for, since the planning and project phase itself takes up to several years.

New Waste Incineration CHP Plant

Municipal waste handling in Frederikshavn is based on the same principles as the rest of Denmark, that is giving priority to recycling of most of the waste, then incineration and only land filling of very small shares. However, the amounts of waste for incineration now exceed the capacity of the two existing plants in the area, and it is planned to build a new. Consequently, the EnergyTown project includes a new waste incineration CHP plant with an expected net-electricity efficiency of 23% and a heat production of 64% and with the capacity of burning 185 GWh/year equal to the available local resources.

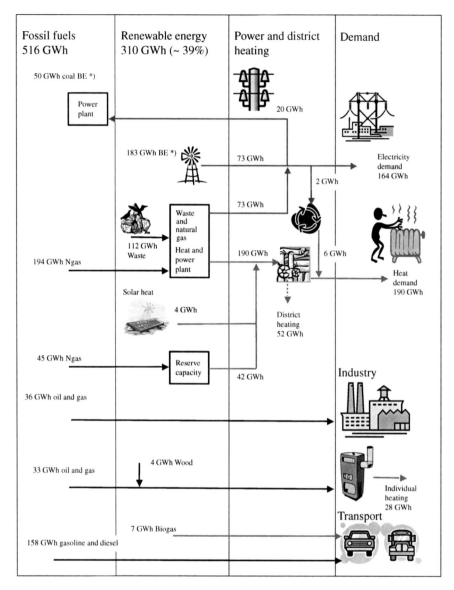

Fig. 11.2   Frederikshavn 2009 (40% renewable energy)

Expansion of District Heating Grid

The project also includes an expansion of the existing district heating grid from the present 190 GWh to a total of 236 GWh. Hereby 70% of the heat demand in industry and individual houses is replaced. The rest of the industry (process heating) will be supplied from biomass boilers, and the individual house heating is converted into a mixture of solar thermal and electric heat pumps.

Transportation

With regard to transportation the project is heading for a solution in which vehicles are converted into the use of biogas in combustion engines (bi-fuel cars), electric cars, and plug-in hybrid cars. In order to implement as much electric driving as possible it is suggested to implement cars which combine the use of batteries with fuel-cell driving based on either methanol or hydrogen. The following specific proposal calculated assumes that motor cycles (4 GWh) and vans and busses (25 GWh) are converted into biogas, hydrogen, or methanol in the ratio 1:1. Of the rest, 10 GWh is converted into biogas 1:1, and out of the rest, 50% is converted into electric driving (1 kWh electricity replaces 3 kWh gasoline) and the rest into FC driving based on replacing 2 kWh of gasoline by 1 kWh of methanol or hydrogen. In total, 165 GWh of gasoline and diesel are replaced by 10 GWh of biogas, 21 GWh of electricity and 61 GWh of methanol.

Biogas Plant and Methanol Production

Partly to be able to produce methanol for transportation and partly to replace natural gas for electricity and heat production, the project includes a biogas plant utilising 34 million ton manure per year for the production of 225 GWh biogas. The facility itself consumes 42 GWh of heat and 7 GWh of electricity.

The biogas can be converted into methanol with an efficiency of 70%. Consequently the production of 61 GWh of methanol is expected to consume 87 GWh of biogas. However, the production of methanol will provide 17 GWh of heat, which can be utilised for district heating.

The methanol may also be fully or partly produced by electrolysis. Moreover, in the end, the cars may consume hydrogen instead of some of the methanol. In such case some of the biogas will be replaced by wind power instead.

Geothermal and Heat Pumps

The town of Frederikshavn is located on top of potential geothermal resources which may be included in the project. The resources can supply hot water with a temperature of approximately 40°C. However, the temperature can be increased to district heating level by the use of an absorption heat pump, which can be supplied with steam from the waste incineration CHP plant. It has been calculated that an input of 13.3 MW steam in combination with a geothermal input of 8.7 MW can produce 22 MW of district heating. The steam input will decrease the electricity production from the CHP plant by only 1.3 MW and the heat production by 11.9 MW. Marginally the absorption heat pump then has a COP of more than 7.

Additional compression heat pumps may be used to utilise the exhaustion gases from the CHP plants and the boilers supplemented by other sources such as wastewater as already included in the plans for 2009. A potential of 10 MWh output is included by use of a heat pump with a COP of 3.

CHP Plant and Boilers

The project includes a biogas CHP plant of 15 MW and efficiencies of 40% electricity and 55% heat. The rest of the heat production will be supplied from a biomass boiler burning straw with an efficiency of 80%.

Wind Power

Finally the project included wind turbines enough to cover rest of the electricity supply, that is, a total of around 40 MW, of which already more than half is going to be implemented by year 2009.

Energy System Analysis

By the use of the EnergyPLAN model some detailed energy system analyses have been conducted of the expected year 2015 system in order to identify the hour-by-hour balances of heat supply and exchange of electricity.

The EnergyPLAN model is a deterministic input/output model. General inputs are demands, renewable energy sources, energy station capacities, costs, and a number of optional different regulation strategies emphasising import/export and excess electricity production. Outputs are energy balances and resulting annual productions, fuel consumption, import/exports of electricity, and total costs including income from the exchange of electricity.

The model can be used to calculate the consequences of operating a given energy system in such way that it meets the set of energy demands of a given year. Different operation strategies can be analysed. Basically, the model distinguishes between technical regulation, that is, identifying the least-fuel-consuming solution, and market-economic regulation, that is, identifying the consequences of operating each station on the electricity market with regard to optimising the business-economic profit. In both situations, most technologies can be actively involved in the regulation. And in both situations, the total costs of the systems can be calculated.

The model includes a large number of traditional technologies, such as power stations, CHP and boilers, as well as energy conversion and technologies used in renewable energy systems, such as heat pumps, electrolysers, and heat, electricity and hydrogen storage technologies including Compressed Air Energy Storage (CAES). The model can also include a number of alternative vehicles, for instance sophisticated technologies such as V2G (Vehicle to grid) in which vehicles may supply the electric grid. Moreover, the model includes various renewable energy sources, such as solar thermal and PV, wind, wave and hydropower.

EnergyPLAN

INPUT

Demands
Electricity
Cooling
District heating
Individual heating
Fuel for industry
Fuel for transport

RES
Wind
Solar Thermal
Photo Voltaic
Geothermal
Hydro Power
Wave

Capacities & efficiencies
Power Plant
Boilers
CHP
Heat Pumps
Electric Boilers
Micro CHP

Storage
Heat storage
Hydrogen storage
Electricity storage
CAES

Transport
Petrol/Diesel Vehicle
Gas Vehicles
Electric Vehicle
V2G
Hydrogen Vehicle
Biofuel Vehicle

Regulation
Technical limitations
Choice of strategy
CEEP strategies
Transmission cap.
External
electricity market

Fuel Cost
Types of fuel
CO2 emission factor
CO2 emission costs
Fuel prices

Cost
Variable Operation
Fixed Operation
Investment
Interest rate

Distribution data

Electricity demand | District heating | Wind | Hydro | Wave | Waste
Solar thermal | Photo Voltaic | Geothermal | Individual heating
Industrial CHP | Transportation | Market prices

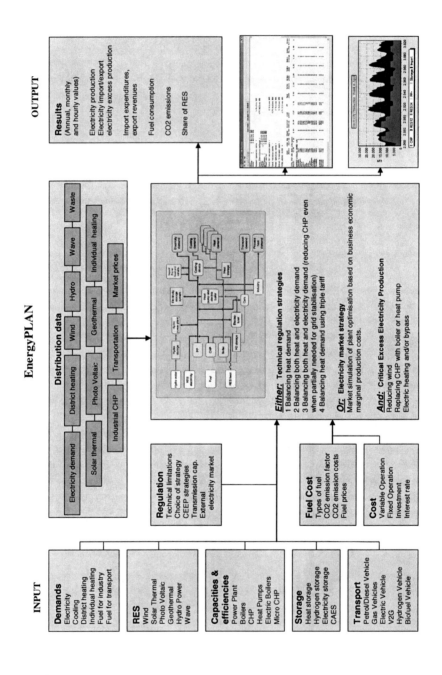

Either: Technical regulation strategies
1 Balancing heat demand
2 Balancing both heat and electricity demand
3 Balancing both heat and electricity demand (reducing CHP even when partially needed for grid stabilisation)
4 Balancing heat demand using triple tariff

Or: Electricity market strategy
Market simulation of plant optimisation based on business economic marginal production costs.

And: Critical Excess Electricity Production
Reducing wind
Replacing CHP with boiler or heat pump
Electric heating and/or bypass

OUTPUT

Results
(Annual, monthly and hourly values)

Electricity production
Electricity import/export
electricity excess production

Import expenditures, export revenues

Fuel consumption

CO2 emissions

Share of RES

The key data in the analyses are hour distributions of demands and fluctuating renewable energy sources as shown in the diagrams. The fluctuations in electricity demand are based on actual measurements of the year 2006 demand of Frederikshavn, and the district heating demand has been based on a typical Danish distribution adjusted by monthly values of the Frederikshavn district heating demand in 2007. Wind power is based on the actual production of the existing wind turbines in 2006. However, the productions have been corrected and adjusted to the expected annual production of an average wind year. For solar collectors, a typical Danish annual production distribution has been used.

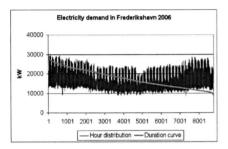

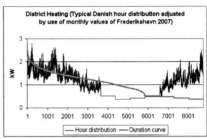

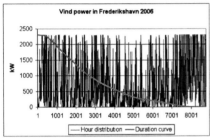

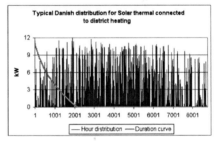

The results of the energy system analyses reveal that if the use of waste for incineration is increased from present 112 GWh to 185 GWh in a plant with the present efficiencies and the present district heating demand, the summer heat production will exceed the demand by 10 GWh. Such excess production will be even higher if heat generated from methanol production is included. However, by building a new plant with higher electric output, expanding the district heating coverage and adding biogas, excess heat production is avoided.

Moreover, the analyses show that on an annual basis all energy demands in the EnergyTown Frederikshavn can be met by 100% renewable energy, that is the use of: 185 GWh waste, 225 GWh biogas, 48 GWh straw, 5 GWh solar thermal, 48 GWh geothermal and 130 GWh wind power (fuel equivalent of 325 GWh). The production of electricity is not able to meet the demands in all hours. The analyses indicate that the system needs an exchange of approximately 25 GWh imported electricity. However, on an annual basis, such import is compensated by a similar export in other hours.

The system is shown in Fig. 11.3.

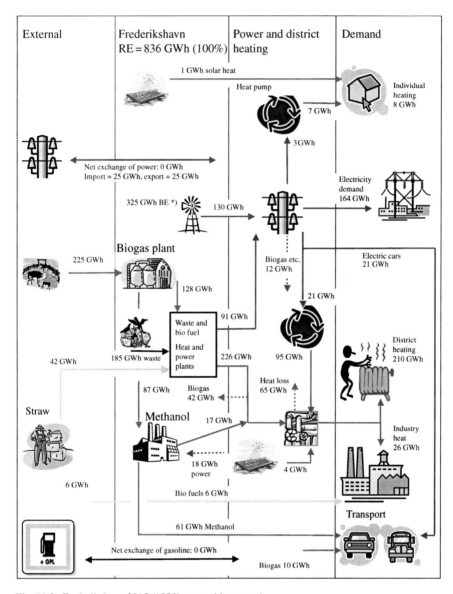

Fig. 11.3 Frederikshavn 2015 (100% renewable energy)

The Third Phase: Frederikshavn in the Year 2030: 100% Renewable Energy and Less Biomass

Phase 3 involves the reduction in biomass including waste to a level corresponding to Frederikshavn's share of the total resources. The potential domestic biomass in Denmark has been identified to 165–400 PJ depending on the scale of biocrops. The residual resources (straw, biogas, wood chips and waste) account for 165 PJ/year. The share of Frederikshavn based on population is then 220–500 GWh/year. The expected year 2015 system utilises around 450 GWh, which will then have to be adjusted accordingly. However, the use of biomass is in the right order of magnitude with regard to the long-term objective of the project.

Phase 3 will, among other things, include a better insulation of homes, power savings and an increased efficiency in the industry as well as further transition to electric cars in transportation. The changes in phase 3 must be coordinated with conditions and activities in the rest of the country. The changes are not made more specific, and no attempt has been made to asses the need for investments.

Climate Change Mitigation in Denmark - A bottom-Up Approach

Frederikshavn represents one out of many cases of towns, cities and areas which make ambitious climate change mitigation plans, involving sustainable or carbon neutral energy systems. In reviewing optimisation criteria for energy systems analyses of renewable energy integration, Østergaard [1] lists a number of such cases at the international level and also elaborates on the sustainable energy system. Moving from the international scene to the regional level, Denmark is one of the places where the level of activity in climate change mitigation is noticeably high. Following the initial steps or working in parallel with Frederikshavn, large cities like Copenhagen, Århus and Aalborg, medium-sized towns like Thisted, Sønderborg and Skive, and islands like Samsö and Lolland are all in the process of designing carbon neutral energy systems. At the national level, the Danish Association of Engineers, IDA, has completed an extensive work on Future Climate, an energy plan which draws on the expertise and voluntary work of more than 1000 of its members. The energy plan is described in detail, e.g., in [2] by Lund and Mathiesen, published in *Energy* - the international journal. In renewable energy strategies for sustainable development, also published in Energy, Lund [3] discusses the perspectives of converting the present Danish energy system into an energy system fuelled only by renewable energy sources.

In fact, the conversion of energy systems into renewable energy systems is a well-researched area; however, this research should not be limited to the technical issue of how to design the systems. The conversion to renewable energy also involves citizens and policy makers who make demands and set forth along new pathways. As argued by Lund [4], the mere awareness of alternatives to conventional fossil-based energy planning is one of the important steps to be taken in a transition towards sustainable energy systems. As demonstrated by the many cases of towns, cities and areas developing sustainable energy plans, this awareness is permeating Danish society. This has the effect that, rather than a government instigating top-down energy planning towards sustainable energy use, communities are actively pursuing goals far more ambitious than the country as a whole. Climate change mitigation is becoming a matter for the general public and, thus, the Danish case represents a bottom-up approach to climate change mitigation.

References

1. Østergaard PA, "Reviewing optimisation criteria for energy systems analyses of renewable energy integration", Energy 34(9): 1236–1245, Sep, 2009.
2. Lund H, Mathiesen BV, "Energy system analysis of 100% renewable energy systems - The case of Denmark in years 2030 and 2050", Energy 34(5): 524–531, May, 2009.
3. Lund H, "Renewable energy strategies for sustainable development", Energy 32(6): 912–919, June, 2007.
4. Lund H, Renewable Energy Systems: The Choice and Modeling of 100% Renewable Solutions. Burlington, MA, USA (Elsevier Academic Press). ISBN: 978-0-12-375028-0.

Chapter 12
Sustainable Communities: The Piedmont Region, Settimo Torinese, Italy

Teresio Asola and Alex Riolfo

Abstract This chapter highlights the projects and achievements dealing with renewable energy and agile energy systems for sustainable communities in Italy.

Both the Piedmontese and Settimo Torinese projects, and other parallel projects in Milan, are regarded as sustainable communities because of their balance between energy consumption and the environment. Additionally, these projects were not only created out of mere respect for the environment, but to create economic value as well.

T. Asola (✉)
PIANETA, Settimo Torinese, Italy
e-mail: teresio.asola@asm-settimo.it

W.W. Clark II (ed.), *Sustainable Communities*, DOI 10.1007/978-1-4419-0219-1_12,   169
© Springer Science+Business Media, LLC 2010

This Chapter

- Presents new energy paradigms in Northern Italy:

  a. Parallel experiences underway: Milano CityLife, Sesto San Giovanni (Falck), Settimo Torinese (Green Tech Park, Laguna Verde)
  b. Settimo Torinese: the R&D archipelago Green Tech Park, a new territory for the experimentation on energy innovation in Piedmont, northern Italy
  c. The use of sustainable, renewable resources and the implementation of new technologies and sustainable practices in Settimo Torinese as a sustainable town

- Introduces "Laguna Verde": a new eco-town of the future in Settimo Torinese, from dream to reality

  d. Global sustainability, resources and technologies in Laguna Verde, Settimo Torinese
  e. Environment/energy balance in Laguna Verde

Innovative Paradigms for Ideas, Projects, and Practices

Energy production must derive from a dialogue with the environment. This seemingly simple concept implies dramatic cultural leaps when considering the design of urban areas as well as the type and amount of energy consumed. Essentially, everything must be reconsidered including how much energy is consumed in buildings, districts, cities, and the territory as a whole, as well as reexamining technologies employed for production and distribution of goods.

Applying new sustainable paradigms requires technical knowledge and design capabilities able to:

- Understand varied energy needs of a specific territory
- Enhance the combination of available resources (wind, sun, water, wastes, biomass, geothermal, etc.)
- Maximize the efficiency of buildings and of the territory as a whole

Settimo Torinese, only a few kilometers from the Piedmont regional capital of Torino, was a typical industrial town. Although it has acquired a definite image of its own, it has always been influenced by Milan, its neighbor to the east. Several railroads cross Settimo Torinese (both new and high speed) as well as several vehicular thoroughfares such as the recently refurbished Torino–Milano highway. Settimo Torinese is an ideal portal toward Milan and a concrete gateway to new eco-city concepts.

Parallel Experiences Underway: Milano CityLife, Sesto San Giovanni (Falck), Settimo Torinese (Green Tech Park, Laguna Verde)

CityLife is a project of redevelopment and requalification of the historic district of Fiera di Milano, signed by Pier Paolo Maggiora (Archa),[1] Zaha Hadid, Arata Isozaki, and Daniel Libeskind according to the guidelines of dialogue in architecture.

CityLife covers more than 250,000 m^2 and is located on one of the main growth thoroughfares of Milano, adjoining one of the most prestigious Milanese institutions, the Fiera. Situated next to several eminent Milanese districts, as well as the ancient network of Navigli waterways, CityLife could eventually prove to be a new city center for Milan.

A major ambition for Milan is to regain a primary role in design experimentation, formal research, and technology innovation. The environmental quality in Milan needs to be improved, which proves to be a remarkable effort. Heavy traffic and air pollution have to be contained in order.

Designing the transformation of the urban areas based on an ecological strategy is the only feasible treatment. Upon specific analysis, the most significant environmental impacts have been pinpointed in order to devote specific care to them as major environmental targets. The project choices have been carried out with particular care to the following items:

- Air
- Water
- Soil and subsoil
- Noise and vibrations
- Car traffic
- Green system and ecosystems
- Urban landscape

The objective of emission reduction due to car traffic is addressed by optimizing the road network and conditions.

Emission reductions are obtained by containing fossil fuel consumptions, along with other solutions adopted for energy efficiency. The solutions combine a use of materials and technological ideas focusing on building insulation, which will be able to guarantee high light transmittance and low thermal transmittance. The materials have been chosen according to their reusability, with a priority to those materials with limited energy needs in the manufacturing cycle.

The technological choices have been oriented to the utmost degree of energy saving and recovery, among them heat pumps (air and water), methane cogeneration (CHP) systems, and photovoltaic systems.

---

[1] http://www.archiportale.com/progettisti/SchedaProgettista.asp?id=87013.

A New District for an Important Milanese Suburban City: Sesto San Giovanni (Formerly Falck Steel Mill)[2]

The requalification of areas in the former Falck steel factory area in Sesto San Giovanni near Milano is an ambitious new urban plan that will transform the town of Sesto, aimed (in Renzo Piano's project) at operating an "urban renewal" of parts of the town of Sesto San Giovanni so far divided by the railway and by the empty lots of the former Falck and Marelli industrial areas. Greenery is one of the distinctive features of Renzo Piano's[3] project.

The whole area, receiving a new population of about 15,000, is reclaimed to ensure the highest health standards, with specific attention to ecosustainability. The area will be provided with an independent energy system adopting a precise strategy based on the use of local resources and the optimization of energy distribution. Groundwater is to be collected and filtered to feed heat pumps, biomasses produced by the Park, and solar power will also be used, leaving the existing cogeneration plant to cover peaks of demand. This system of optimization of energy production and distribution plans to use trigeneration plants for the simultaneous production of electricity as well as heat and air conditioning and tunnels feeding all the blocks along the main streets.

PIANETA[4] (belonging to the utility group, Settimo-based ASM) has tightly cooperated with Buro Happold Ltd[5] to help create an energy masterplan of the new urban area project by Renzo Piano Building Workshop (RPBW) and the Physics Nobel Prize–recipient Prof. Carlo Rubbia.[6] The aspects related to urban planning and construction were assessed by Risanamento SpA.

The heart of the old industrial town is to be rebuilt based on a new, sustainable model, in which energies are produced in several sites and in several ways, and distributed by means of a network aimed at balancing demand and offer. The incoming 40,000 inhabitants will have energy consumption per person estimated to be nearly half as much as the other Milano inhabitants. Consequently, the resulting new city is bound to grow more rapidly than ever, because both people and companies will prefer a cleaner, more efficient, and cost-effective city to live and work in. The energy masterplan envisions an on-site production of a major quota (more than 75%) of the energy needs (electricity, heating, cooling) by means of high-performing, agile, small-scale energy systems, mainly based on the use of locally available renewable resources. The project aims at producing energy by maximizing resources and overall efficiency and minimizing the environmental impact, noise, and emissions.

---

[2] http://www.risanamentospa.it/upload/nenergiaOtt06.pdf; http://www.risanamentospa.it/web/start. asp?idLingua=2.

[3] http://rpbw.r.ui-pro.com.

[4] http://www.pianeta.eu/ITA/sesto.html.

[5] http://www.burohappold.com/bh/home.aspx.

[6] http://nobelprize.org/nobel_prizes/physics/laureates/1984/rubbia-autobio.html.

The design of the overall energy system has been developed on the basis of the following objectives (generating from such keywords as flexibility, independence, modularity, and innovation), aimed at developing innovative energy solutions in order to help meet modern standards of efficiency and savings required by law and by international protocols:

- Minimizing the environmental, dimensional, and economic impact of the installations.
- Energy independence and self-sufficiency, but at the same time compatibility and integration with existing neighboring urban areas infrastructures (cogeneration plant, district heating).
- Use of all existing resources within the site, both traditional and alternative/renewable.
- Efficiency and effectiveness of energy systems of production, distribution, and use of energy.
- Reduction in consumption of fossil fuels and polluting emissions, as well as a commitment to continue researching innovative solutions.
- Flexibility and adaptability to fast-changing technologies.

The method and the solutions used to address the above objectives have been as follows:

- Implementation of good sustainable design principles.
- Efficient use of on-site resources, most notably groundwater.
- Not an exclusive energy solution for the site; rather a flexible mix of systems and scales.
- Groundwater extraction and heat pumps for heating and cooling of buildings.
- Extension of existing district heating system for warm water in buildings.
- New trigeneration energy centers for electricity generation, heating, and cooling.
- Building basement technology rooms to minimize obtrusive roof-mounted plant.
- Good passive design of buildings to reduce energy consumption.
- Integration of alternative energy and renewable energy systems.
- Sustainable mobility.
- Environmental assessment schemes such as the Italian ITACA or US LEED to measure and score environmental performance for specific buildings.

As to energy production, there is no exclusive, single solution for the site, but rather a mix of two or three main primary energy strategies. The energy strategy focuses on combining the most efficient use of existing opportunities (e.g., district heating and the cogeneration plant) with the use of distributed generation systems, mainly trigeneration plants and alternative energies, particularly groundwater coupled with heat pumps. An efficient use of resources will reduce both energy consumption and noxious emissions, the target being to reduce carbon dioxide emissions by 50%, compared to conventional heating, cooling, and electrical systems.

Alternative energy systems can be implemented on a small scale as a demonstration of new technology matched to a specific use or on a larger scale integrated into buildings. Solar systems mainly concern conventional solar thermal collectors, high-intensity solar thermal collectors, and photovoltaic solar cells. Conventional solar thermal collectors can heat water up to 80–100°C for use in sanitary water systems for washing or swimming pool heating. PVs generate electricity to export to the electrical distribution grid to balance demand for street lighting. High-intensity solar thermal collectors using parabolic mirrors to intensely focus solar radiation onto oil-filled tubes can heat the oil to very high temperatures (500–600°C) and then generate electricity by means of steam turbines.

Biomass: Focus has been made on energy production out of biomass and organic waste from park maintenance "products" (pruning, chopping): drying and gasification allow for excellent energy recovery, as well as providing a viable solution to the problem of disposing of organic substances. In the longer term, as the landscape matures, it is possible to take wood off-cuts from trees for use in wood burning boilers to generate hot water for heating or to be utilized in a gasification or pyrolysis process to produce energy.

Hydrogen: Hydrogen is produced from either water in electrolyzer using electricity or methane (in natural gas) in a reformer. Fuel cell technology is still developing and its possible applications range from mobility to stationary. An interesting application for the site is to use electricity from PV solar cells to generate "green" hydrogen and oxygen from water in an electrolyzer. The oxygen can be used directly for purification of the groundwater, and the hydrogen combined with air, in a fuel cell to regenerate electricity.

Hydrogen will be used to experiment sustainable mobility in the new district of Sesto based on the use of hydrogen as fuel for motor vehicles (SHMS, Sesto Hydrogen Mobility System). The project envisages the creation of filling stations (SHRS Sesto Hydrogen Refuelling Station) and the means of transport[7] (buses and scooters like the Torino Environment Park's one) based on fuel cell technology.

The objective of the SHRS is to produce fuel (hydrogen) mainly for local public transport. The SHRS is implemented using the state-of-the art production and storage technologies. The supply section in the Falck project includes a 350-bar dispenser for the hydrogen supply of buses.

Hydrogen produced out of water electrolysis, like in two paradigm projects,[8] was realized by PIANETA in Settimo Torinese in 2005 for an ASM office building and in Cesana for the Winter Olympic Games ("Primo Settimo" and "HighHy", see box below), in tight cooperation with the Regional Government of Piemonte.

---

[7] http://www.iaad.it/eng/gallery/gall_07/FormulaHYSY_gara_eng/index.html;
http://www.envipark.com/index.php?option=com_frontpageItemid=1.

[8] http://www.pianeta.eu/ITA/primo_settimo.html; http://www.pianeta.eu/ITA/h2006.html.

In "Primo Settimo," the first integrated energy plant in Italy designed for the production, storage, and use of "green" hydrogen in an office building, the hydrogen power plant is utilized to give energy to the parking lot of ASM office building.

Electricity is produced by 2 PEM hydrogen fuel cells. Hydrogen is produced during day hours by an alkaline electrolyzer (utilizing a green 70 KW photovoltaic system) able to generate 200-bar hydrogen to be then stored.

The target of Primo Settimo is to store energy produced during day hours by electrolysis transformation into hydrogen and to utilize it at night hours, in order to light the parking lot. Production and utilization of hydrogen are not contemporary: during the day hydrogen is produced; at night the fuel cells operate. This system was created to generate clean energy for stationary use and, at the same time, to supply hydrogen for future fuel cell vehicles. The system was designed by PIANETA and ASM to also become an open-air laboratory for research activities for innovative technological solutions to be applied to the hydrogen supply chain (production, storage, and use), together with Politecnico di Torino Polytechnic University and Regione Piemonte.

"HighHy," a smaller-scale hydrogen stationary energy production plant, was developed by PIANETA and ASM for the 2006 Torino Winter Olympic Games on the same basis as "Primo Settimo," in the "Olympic mountains" of Cesana Torinese.

Settimo Torinese: The R&D Archipelago Green Tech Park, a New Territory for the Experimentation of Energy Innovation in Piedmont, Northern Italy

Piemonte, in northwestern Italy, borders Switzerland to the north, France to the west, the valley of Aosta to the northeast, and Lombardia (whose main city is Milano) to the east. To the southeast it borders Emilia Romagna, and Liguria to the south. The Piedmont government has been investing a lot on the new challenges of clean, sustainable energy. In particular, on May 24, 2008, the president of the Piedmont region, Mercedes Bresso, announced the beginning of the "Piedmontese energy oil independence war," promoting and signing the "manifesto for the petrol energy independence."[9] Said manifesto, begins with a proposition that states, "in order to reduce energy dependence from fossil fuels, it takes an extraordinary collective, daily commitment," and points out that "Piedmont has decided to accept the challenge by investing on renewable energies, energy saving and on sustainable technologies in order to secure a future to our sons, with a collective commitment

---

[9]http://www.regione.piemonte.it/energia/images/stories/dwd/manifesto.pdf.

of the whole population." In order to concretely become "the Italian ecological engine" a concrete, everyday commitment must be taken by everyone from now to 2020 in order to quickly abide by the EU targets (aimed at maximizing production from renewable sources, enhancing energy saving and production of biofuels not derived from food sources but out of cellulose and wood, and dramatically curbing the greenhouse gas emission). The goals pointed out by Regione Piemonte as mandatory ways to achieve the European targets are listed below:

- To enhance renewable energy sources and overall energy saving.
- To design houses and workplaces able to self-produce heat and electricity, granting comfort and respect for the environment.
- To support research and enhance energy savings and self-production.
- To use public means of transport, as well as highly efficient, nonpolluting vehicles.
- To support an agricultural production looking after ecological values and the resources locally available.
- To upbring the younger generations toward a more sober, rational, and fair energy culture.
- To give everyone a real chance both to draw on as well as to input energy by an open and widespread grid network, in which everyone is a producer and a consumer at the same time.

In Piedmont, particularly in some areas of Torino and in the suburban territory immediately northeast, there are already initiatives and realizations to be accounted for. The projects devoted to innovation and agile energy systems aimed at distributed generation are as follows:

- Energy recovery out of biogas in an AMIAT[10] garbage dump in Torino and in a SMAT[11] purifier site (the largest chemical–physical–biological treatment system of waters in Italy).
- Small-scale hydroelectrical and PV systems in the Environment Park[12] building.
- Medium-sized cogeneration plants in Torino, Settimo Torinese, and Chivasso as well as district-heating grids, both realized by SEI (ASM Group).
- Medium- to large-scale fuel cells production by Siemens[13] in Torino.

---

[10]Company for the complete waste treatment, disposal and recovery cycle; http://www.amiat. it/pagine.cfm?SEZ_ID=23

[11]Company leader in the field of integrated water services; http://www.smatorino. it/area_istituzionale_eng?id=1.

[12]It provides innovative solutions in the fields of energy/environment; http://www.envipark. com/index.php?lang=en.

[13]http://webdoc.siemens.it/CP/PW/ProdottieSoluzioni/PowerGeneration/CelleaCombustibile/ default.htm.

- Medium-scale fuel cells production by EPS Electro Power Systems[14] near Torino.
- PV plant[15] for gas boxes, realized by PIANETA.
- Three-generation system with a PV system and microturbines integrating the first plant for production, storage, and use of hydrogen in Italy, in Settimo Torinese, realized by PIANETA for ASM Settimo Torinese[16]
- Microturbines, as well as other realizations, mainly in Settimo Torinese.

Within this framework, the Settimo Torinese town administration, a Piedmontese town with a population of 50,000, adjoining Torino, and 80 miles from Milano, has been actively and proactively working for at least the last 5 years.

In the suburban area northeast of Torino, an innovative energetic pattern has been developed, actively promoted by Aldo Corgiat Loia,[17] mayor of Settimo Torinese – a pattern arising from experiences and vocations existing in the territory, but not thoroughly structured yet before: the one regarding the renewable sources and the agile energy systems.

The model for the energy innovation network is implemented on a wide suburban area surrounding Torino pivoted on Settimo Torinese. A model was developed by PIANETA with many municipalities, with the Municipality of Settimo Torinese and Mayor Aldo Corgiat Loia as a leader and promoter, called "Green Tech Park." The target of the Green Tech Park is the implementation of new energy paradigms, starting from the territory, local resources, existing cultures, and experiences.

The pattern being developed on the wide area is the energy web: a network of sustainable energy paradigms, differentiated by technology and by application, where each and every knot of the network is able (like the Internet, peer-to-peer) to be a producer and a user at the same time, all of it on an environmentally correct background (a new park "Grande Parco 2011" crossing the towns being part of the Clean-Tech Park [CTP]), and to get more momentum out of its proximity with the Olympic city Torino, a city able to guarantee a proper welcome to possible visitors. The CTP is growing as an "open-air" laboratory and living energy museum.

Energy is seen as a concrete development factor and not as a mere environmental opportunity – an entrepreneurial development too and consequently, again, a renewed territorial plan. Energy as an occasion to match the development needs of economy with the environmental sustainability of territory. Energy opportunities are

---

[14]Company for the production of fuel cells running on pure hydrogen, natural gas, LPG, and ethanol, for backup and stationary use in general. Based in Torino (Italy), it is an established player in the fuel cell systems sector having developed and commercialized the Electro7™ power system. Electro7™ is the first multi-output fuel cell system for business continuity applications and provides 100% clean power with virtually unlimited autonomy. www.electrops.it.

[15]http://www.pianeta.eu/ITA/sunjump.html.

[16]http://www.pianeta.eu/ITA/primo_settimo.html.

[17]Mayor of Settimo Torinese since 2005, formerly chairman of the Utility Company ASM Settimo.

designed and implemented in each and every form on the whole surface of the territory, in the form of an "archipelago-like" technological park (Green Tech Park) characterized by a manifold system of production "isles." In the Green Tech Park a rich open-air catalogue grouping the newest technologies – as far as renewable and new "intelligent" forms of energy are concerned – is experimented and permanently displayed and exhibited, in order to further nurture the "knowledge society." The use of resources locally available must be made on the basis of the capacity to integrate different innovative technologies into a network able to guarantee their overall governance and control.

One of the main objectives of the Green Tech Park (CTP) project led by Settimo Torinese on the suburban area is to create a widespread but consistent and coherent set of sustainable energy realizations based on the integration of different, highly innovative paradigms of energy production, distribution, and management.

A few examples of a set of energy "isles," good practices of energy efficiency and innovation, and integrating technologies are as follows:

– Photovoltaic systems, thermal solar systems
– Cogeneration (microturbines, natural gas endothermic engines, hydrogen fuel cells)
– Mini-/micro-wind systems
– Ground/water heat pumps
– Small hydroelectric systems, geothermal systems for electricity
– Waste/biomass gasifiers
– Fuel cell–based backup systems, stationary hydrogen plants

The Use of Sustainable, Renewable Resources and the Implementation of New Technologies and Sustainable Practices in Settimo Torinese as a Sustainable Town

Some of the planned, underway, or realized projects in Settimo Torinese are as follows:

• Recovery of energy out of fragmented waste treatment and treatment and recovery of energy out of urban solid waste, biogas, organic waste, biomasses, photovoltaic plants (small-scale to big PV plants), heat pumps, small hydro systems, and mini-wind systems
• New urban blocks and factories designed according to bioarchitectural principles, energy/environmental sustainability
• Low-cost ecofriendly new houses, which use active and passive technology, in order to transform the building into a bioclimatic engine
• Valorization of public buildings, according to the principles of ecosustainability (bioarchitecture and production plant efficiency)

- New regulatory systems: innovation in the sets of rules concerning the building and planning systems and energy certifications

In this way, the Municipality of Settimo Torinese is leading other neighboring municipalities to follow suit. It is bound to become an incubator of new initiatives (both public and private) and of case histories connected with eco-energies and the ecosustainability of buildings and towns: the creation of a new energy culture based on local experiences.

A word must be spared on the project of a new set of rules, being approved by the Municipality of Settimo Torinese and by other neighboring municipalities, aimed at enhancing good, ecological, energy-sensible practices: a regulatory document dealing with the requirements necessary for the "eco-energetical" and "bio-energetical" certification of buildings, as well as with incentives to be given out to good realized practices (both active and passive technologies).

PIANETA (see box below), an energy service company based in Settimo, has become a major ingredient of the "sustainability recipe" in Piedmont (in tight cooperation with the municipalities) – the engine for the development of challenging projects aimed at sustainable communities.

PIANETA (www.pianeta.eu) is an all-round energy-saving company, focused on ecosustainability of buildings, districts, and cities: energy self-sufficient, cost-effective buildings, districts, and cities

As the cook turns to ingredients of the territory, so PIANETA prepares its own energy production systems and passive solutions basing its efforts on the renewable energy resources that the territory offers.

PIANETA is entirely focused on design, implementation, and management of agile energy systems based on global sustainability and all-round efficiency. An energy system is "agile" when it can satisfy several energy needs in different sites, utilizing the resources immediately at hand on-site and the optimized mix of high-efficiency energy technologies. The agile energy systems of PIANETA provide different quantities of energy (from kilowatts to megawatts of electricity, heat, and cool) using natural, renewable sources (sun, wind, water, geothermic) locally available. PIANETA's agile energy systems grant self-sufficiency to buildings, housing and commercial estates, industries, new city districts, new cities, and whatever urban area seeking cleaner and more efficient energy solutions. PIANETA produces energy where and when necessary, wastelessly, efficiently, effectively, and in the amount needed: a company specially devoted to design, realize, and manage small, innovative energy plants as compared to the traditional, big energy plants, integrating the most innovative technologies for the employment of the best, sustainable source to fit in the project. Right from the beginning of its existence, PIANETA has contributed to bridge the gap between university labs and the market, in many fields of renewable energies, being the first in Italy, for instance, to design and implement plants for H_2 out of renewable resources (Primo Settimo and HighHy, which have contributed to define the Torino 2006 Winter Olympics as "the first hydrogen Olympic Games").

Furthermore, PIANETA has been working on the assumptions that

- architecture must save natural resources (water), regenerate natural resources (O_2), and minimize CO_2 production; a building basically mustn't consume energy; it must produce it, utilizing renewable sources (sun, water, wind, motion) and minimizing wastes (to be used themselves as an energy source);
- the best green, clean energy is the one we neither produce nor use
- energy efficiency is a synthesis between sustainable production and its sustainable use and between active and passive technologies
- the building is a machine that shouldn't consume resources and energy, but rather saves and produces

"Laguna Verde": A New Eco-Town of the Future in Settimo Torinese, from Dream to Reality

A very important example of a global sustainability project, consistent with (and inside) the Settimo Torinese Green Tech Park, is a new, daring, greenfield development district of Settimo Torinese (a new area, a little eco-city – formerly an industrial district in Settimo Torinese – nearly as wide as 1 million m^2 in a suburban area of Settimo, some 10,000 fresh inhabitants, university, high-tech R&D Companies) called "Laguna Verde" (Green Lagoon).

Innovation, environment sustainability, project, and architectural dialogue are the main elements that make this new urban centrality in Settimo one of the most important requalification projects.

Laguna Verde

Settimo Torinese Mayor Aldo Corgiat Loia and architect Pier Paolo Maggiora created the guidelines for Laguna Verde according to their vision of a sustainable community (see Green Park). The design was outlined by Pier Paolo Maggiora's ArchA, along with the cooperation of eminent experts in various fields of sustainability. The challenge was to create a new concept of a town not merely out of an architect's pen, but instead out of a dialogue realized by a heterogeneous yet coherent team of specialists of different topics, bearers of different cultures and experiences.

Fig. 12.1 Laguna Verde

The sustainable energy project will be characterized by the newest and most innovative technologies for "passive" building performances and energy production, as well as resource savings, water reuse, photovoltaic and microeolic systems, "green" H_2 production, and agile energy systems.

The total surface area covers one million m^2 and, at the end of the operation, will offer 320.000 m^2 of public green space, a 60.000 m^2 "island" for R&D and higher education, sports facilities, a swimming pool and a museum (about 55 m^2), as well as a high-level, innovative energy-efficient housing estate.

Structural Details

The structure itself explains the name Laguna Verde (Green Lagoon). Some compact islands will be developed (16 just for the residential part); they will emerge from a lagoon made of parks and gardens below: a "natural," horizontal level of greenery (integrated with an artificial vertical garden and parking lots with technological plants hidden inside) in continuity with "Tangenziale Verde"[2] (Green bypass) north of Torino, a green area extending on several million meter squares, with the Torino Hill Park leading to Superga and the river Po.

All the buildings will be elevated from the ground and connected and highlighted by a large suspended central avenue fraught with public and private functions so that the existing greenery below can live alongside the structures instead of being interrupted and removed.

The investment will exceed 1.3 billion euros (net worth of the property) and is expected to rise by another 20–40%, thanks to the solutions chosen to obtain the maximum environmental sustainability.

In Settimo Torinese, the Laguna Verde project will transform a significant portion of the city by radically transforming the existing district into a futuristic eco-city. The eco-city will be totally elevated and characterized by tall towers and an overall care for the environment. This is possible, thanks to a concrete cooperation between public owners (municipality) and private owners (manufacturers and real estate companies) in the creation of a newly registered company.

Innovations have also taken place on the administrative level as a result of the size of the transformation. Both the municipality and the private enterprises emphasized their ambition to do so, as a project of this size with such limited time schedule has so far been unparalleled in Italy.

From dream to vision to reality, Laguna Verde in Settimo Torinese is becoming more real every day. Progress is already visible in a big portion of the city, thanks to the quick and affective approval by the Settimo Torinese city council.

Global Sustainability, Resources, and Technologies in Laguna Verde, Settimo Torinese

Laguna Verde in Settimo Torinese is one of the most ambitious projects of urban renovation, in terms of both volume and conceptual difficulty, in Italy. It is developed on a surface of about $1\,km^2$ along one of the main Settimo avenues, Via Torino. This area is home to many public and private properties such as a tire production plant, which will be relocated (and is bound to become the most innovative tire production plant in the world) to another district in Settimo Torinese itself.

Laguna Verde was inspired by the principle of global sustainability and confirmed a multidimensional vision and approach. ArchA experts break down sustainability concepts into five main issues that must all interact:

1. Natural-ecological and environmental sustainability, defining the best equilibrium of relationships of natural environmental spaces (as well as resources: water, greenery, soil) with the spaces of human-built artificial environment in order to guarantee a compliment between human and nonhuman, inner and outer environments.
2. Anthropological/cultural sustainability, that is, the dynamic continuity between culture and habits through well-balanced growth processes and a gradual progressive change. The change is needed in order to define the identity of emerging open cities that are connected with the world while being wisely deep rooted in their own territory.
3. Sociological/management sustainability, or the governance of the city in order to favor qualified and highly skilled government practices.

4. Architectural/urban planning sustainability: the quest for a perceptive harmony, overall consistency, ability to recognize, accuracy, and style.
5. Economic/financial sustainability: concerning the necessary investments for the start-up as well as all the resources the new city can acquire, thanks to its overall attractiveness. The new city will attract investments, enterprise settlements, new services, and the growth of new productive and service activities with the consequent generation of wealth and added value.

Laguna Verde is defined through a pattern of urban structure characterize by a high overall sustainability. The concept designed by Pier Paolo Maggiora and his team of experts takes up the idea of open source and global architectural dialogue.

The actual structure involves the clustering of built areas of compact blocks. These blocks are essentially islands surfacing in a green lagoon and are made up of parks, gardens, and green areas (two-thirds of the whole area is available for greening).

In the "public city" sector, it has been planned that a whole "island" should be devoted to university research and higher education.

The research project developed by the Polytechnic University of Torino, Politecnico, is certainly one of Laguna Verde's main structural strengths. In terms of architecture and technology, Laguna Verde is a new urban sight into the future. The idea of research derives mainly from recent agreements signed with Pirelli for the development of innovative production technologies for "intelligent tires" and aims to create a polytechnic citadel based on keywords like education, research, and industry, with the university and municipality as primary players.

According to the Dean Francesco Profumo of Politecnico of Tornia, "considering the situation of fossil fuels, whose costs can stop rising, it is more and more strategic to focus on R&D and the realization of sustainable processes and products. Accordingly, and important project like Laguna Verde can't be conceived without being based on environmental sustainability and eco-friendliness."

The ecological city has to rethink its three crucial subsystems: production of goods and services, construction, and mobility. Every subsystem must be able to cooperate with others, so as to utilize its own wastes as active inputs, transforming them into useful inputs for all components of the system. The objective is to give a concrete distinctive sign to the entire Torino metropolitan area characterized by a high relationship density, similar to Milano and Geneva. Laguna Verde physically and conceptually links Torino with Settimo Torinese, as well as Torino with Milano, assigning a new value to the anthropic settlement, a value which is not based on gross floor service or volumetric index of area, but rather on urban and building quality and to correct energy management of the new area, a district meant as a non-energivore, highly efficient organism.

Laguna Verde will use a wide range of renewable sources as a distinctive trait, along with the rational use of energy and an intelligent management of energy demands. A 10.000 people district (eco-city) like Laguna Verde, if based on traditional energy habits, can produce as much as 32 tons of CO_2 per year, as well as hundreds of tons of pollutants in gaseous, liquid, or solid state on a relatively small territory. Besides the consequences on quality of life, it is possible to directly interrelate such emission levels with public health expenditure, which increases exponentially in urban centers more densely populated.

The design of the eco-city aims at planning not only the infrastructure/building system, but also the development of enough greenery to act as a "lung" to absorb CO_2 and thin dusts.

Laguna Verde utilizes hydrogen-powered motors for public transport (integrated mini fuel cells for buses, shuttles, etc.), electricity-powered motors fueled by photovoltaic energy from systems integrated in infrastructures and in vehicles themselves, and "archiwind" systems homogeneously distributed in the eco-city where small-scale aerogenerators are integrated in the means of transport as well as in the streets, in order to obtain a constant balance between produced and demanded energy for mobility; 75% of energy needed for transport is renewable.

The kinetic energy out of motion and pulses produced by the vibrations of vehicles is captured and transformed into electric energy by means of a dense network of dynamo-switching sensors. These are properly distributed on the moving structures, which are able to transmit the captured energy, and switch to conversion and rectifying systems and then to the electric grid.

Fig. 12.2 Geographical image of surrounding region

Energy Production and Consumption

Every resident in Laguna Verde is a producer of his own needed energy: a little, but valuable engine of the new urban energy economy. The electric grid and the district-heating network play the role of giant accumulators, nurtured by the users themselves; the residents and their building become the main local energy producers. The TS/C (transmitting surface/conditioning) ratio of buildings is thoroughly studied for that climatic zone, with slight adjustments based on the size of buildings and the use they are zoned for. The shape of buildings arise from their zoned use, their function in the general settlement organism, the natural peculiarities of the surrounding environment, as well as, first of all, the climate conditions in that specific area.

– Every residential/service building-plant system doesn't consume more than 25 $KWh/m^2/year$ (PEYD Primary Energy Yearly Demand).
– Every factory's building-plant system (excluding productive process energy) doesn't consume more than 35 $KWh/m^2/year$ (PEYD Primary Energy Yearly Demand).
– Every residential/service building produces between 25 and 40 $KWh/m^2/year$ of electricity from renewable sources (primarily sun and wind), from treatment of wastes (small-scale gasifiers and batch gasifier systems or plasma converters), from small cogeneration and trigeneration, and as far as thermal energy is concerned, latent heat/cool accumulators, etc.).
– Every residential/service building emits 90% less CO_2 and other greenhouse gases as well as 90% less thin dusts and noxious nanoparticles.

Water

At an urban level, a sustainable, ecofriendly management of water aims at eliminating tapping into deep-level underground water, in order to preserve the quality and to guarantee its equilibrium. This is a fundamental step toward the reduction of subsidence and the preservation of a water reserve useful for the supplies. This implies a recourse to the use of surface waters, taking place by restoring and ensuring a minimum vital flow in rivers and in the secondary waterways, a condition to safeguard the water life, as well as to guarantee a refill of the underground waters. As to waterproofing, particular attention has been devoted to compensate the effects of building. Instead of collecting rainwater directly into the sewage system or surface waterways, a design has been created to install a system of catch-basins of rainwater-collecting loglines. This is modeled after the natural process of rainwater collection ensuring the regeneration of groundwater. Every residential/service building or factory picks the meteoric water, selects the first rainwater in controlled pools, and recovers it for non-drinking uses. The first rainwater undergoes treatments of desilting and filtering and is recovered for watering and washing chores; 100% of water used by Laguna Verde is recovered after its use.

Everything in the new eco-city including the shape of residential and service sector building units as well as the development of transport lines and so on arises from the on-site energy flows, in order to eliminate the creation of urban heat islands.

Wind

The morphology of urban settlements will harness the wind directions inward, pushing toward the points of exploitation.

Small energy systems, aimed at a complete integration and "fusion" into the architectural elements, will be employed. For instance, tiny wind turbines in recycled, light, and cheap plastic whose diameter is around 25 cm are placed in sequence to produce small quantities of electricity which can satisfy the home demand for green, clean energy, at zero cost. Advantages include the fact that such systems get actuated by winds as weak as 2 m/s and keep working for 80% of the time. Compared to a traditional alloy mini-wind system, the plastic type has an 80–90% lower production cost.

Solar

In Laguna Verde, new experiments aim at a concrete lowering of the production costs, experimenting some areas' photovoltaic silicon-free technologies, and relying on new materials and new methods for the production of electricity. In the buildings, new solar technologies will be employed, jointly with traditional silicon-based ones. A third generation of solar cells, completely integrated into the building structures, are obtained through print on thin, flexible material of nanoparticles in an alloy of copper, indium, gallium, and selenium (CIGS: their efficiency is the same as the silicon ones, but their cost adds up to about a fifth); this technology can take the production cost of electricity as much down as to 99 cents. This is less than fossil fuels like coal and thin films now on market (whose cost is around 3.55 USD/W). Integrated in the "intelligent skins" of the buildings, Laguna Verde will experiment technologies normally used on solar-powered space stations, such as multi-junction solar cells combined with a solar concentrator under the name of CPV. In these kinds of cells, the concentrator has the size of $1 \, cm^2$ and is able to generate the same amount of energy as $500 \, cm^2$ of traditional solar cells. This new technology can greatly reduce the production cost of electricity out of renewables and is a fundamental step toward the achievement of one of the main objectives of the project "High Performance PV" by the US Department of Energy, that is to say the development of solar cells able to turn more than 33% of the sun energy into electricity.

CPV or XCPV technologies can be completely integrated into the architectural shape. They are designed to let the sunlight pass through the connection joint between the modules and between the single concentration cells. In most cases, modules are matched with exchangers, in order to absorb the excess heat. The microsystems are activated by means of sensors detecting the temperature of the inner face of the modules. When it exceeds 45°C, a junction box actuates micro-exchangers in order to lower the temperature within limits and to recover the excess thermal energy. Therefore, by keeping the concentration modules' temperature within the optimal values, it is granted the constant utmost efficiency of the system. Furthermore a "cool" layer gets created, which, along with the dehumidifiers, helps to alleviate the need for air conditioning.

Geothermal

Consequently, geothermal systems will use the free, renewable energy available to save up to 70% on heating, cooling, and hot water. Geothermal heat pumps (GHPs) (sometimes referred to as GeoExchange, earth-coupled, ground-source, or water-source heat pumps) have been in use since the late 1940s. GHPs use the constant temperature of the earth as the exchange medium instead of the outside air temperature. This allows the system to reach fairly high efficiencies (300–600%) on the coldest of winter nights, compared to 175–250% for air-source heat pumps on cool days. While we normally experience seasonal temperature extremes – from scorching heat in the summer to sub-zero cold in the winter – a few feet below the earth's surface the ground remains at a relatively constant temperature. Depending on latitude, ground temperatures range from 45°F (7°C) to 75°F (21°C). Like a cave, this ground temperature is warmer than the air above it during the winter and cooler than the air in the summer. The GHP takes advantage of this by exchanging heat with the earth through a ground heat exchanger. As with any heat pump, geothermal and water-source heat pumps are able to heat, cool, and, if so equipped, supply the house with hot water. Some models of geothermal systems are available with two-speed compressors and variable fans for more comfort and energy savings. Compared to air-source heat pumps, they are quieter, last longer, need little maintenance, and do not depend on the temperature of the outside air. A dual-source heat pump combines an air-source heat pump with a GHP. These appliances combine the best of both systems. Dual-source heat pumps have higher efficiency ratings than air-source units, but are not as efficient as geothermal units. The main advantage of dual-source systems is that they cost much less to install than a single geothermal unit and work almost as well.

The building shell, as well as shingles and sunscreening/shading systems, will not play the role only of the skin of the architectural organism, but also of an energetic interface through the integration of systems for the production and the distribution of energy.

Other Remarks

The town of the future, Laguna Verde eco-city in Settimo Torinese, goes along with natural resources and uses them without abusing them. Laguna Verde is meant to be a living organism. Urban greenery must be read in the new eco-city as an "ecological complement of the building." The areas underneath the buildings are characterized by widespread greenery illuminated by an innovative system of sunlight transport through micro-solar tubes. The ecofriendly design approach aims at the improvement of the global ecological quality by means of actions like microclimatic regulation, integration between natural elements and built elements, as well as the production of work and of ecofriendly economies.

Laguna Verde in Settimo Torinese realizes new inner ecosystems through the creation of horizontal and vertical phytopurifying areas, which favor biodiversity as well as water recovery, green areas at the foot of Laguna Verde, and decorative green vertical areas integrated with intelligent skins in tall buildings, for the creation of passive greenhouses and vertical winter gardens.

Laguna Verde in Settimo Torinese is bound to become the town of recycling and of global sustainability: Laguna Verde, as said, transforms local natural sources (sun, wind, motion, wastes, thermal differential) into electricity, hydrogen, gas, and heat, matching active technologies with passive building organisms, so as to make up a self-sufficient, low-emitting, 100% sustainable system.

Waste Management

Laguna Verde can draw a concrete advantage out of waste as well. Laguna Verde will only produce only 25,000 tons/year of urban solid wastes in residential areas. A recycling/reuse system for urban solid wastes grants a reuse of up to 70% of the waste produced by means of home separators integrated in the building, as well as small-scale systems for the treatment of human waste and energy production systems out of all the remaining wastes. It is then possible to produce syngas (and therefore energy out of it) from waste, in every building or block, aiming at the utilization of 100% of unrecyclable wastes.

As the price of oil and other energy resources rises, waste products, normally considered as mere costs and a nuisance, have become a valuable resource that can no longer be ignored. Energy from waste not only reduces our reliance on fossil fuels such as coal, oil, or gas, but also reduces the amount of land used for landfills. Waste can be economically used to produce heat, electricity, and other usable energy forms for industrial and domestic use.

Every district/housing system will be equipped with a mini-plant for waste disposal/treatment, whose technology, rather than based solely on incineration, aims at experimenting new energy production processes.

The recovery of energy from waste materials after their gasification is already a well-established technology. The most promising use for waste-produced energy is in the generation of electricity. High-pressure steam can be used to drive a turbine in order to generate electrical power, an efficient and well-proven technique.

The solutions for energy recovery from the syngas largely depend on the kind of energy needed:

Burn the syngas for heat, 95% thermal 0% electricity
Burn the syngas for steam turbine, 70% thermal 15% electricity
Engine for electricity, 50% thermal 37% electricity
Fuel cell, 35% thermal 55% electricity

The amount of energy that can be recovered from waste depends on the type of waste, its moisture degree, and the caloric (BTU) value. For example, a plant with 60 tons of typical MSW (municipal solid waste) per day is capable of producing 1–2 MW of electricity or 4–8 tons per hour of steam. With industrial and bio-waste or oily absorbent waste and tires, which are drier and have a higher caloric value, the value of the recovered energy may be significantly higher.

Compared with traditional incinerators, the energy efficiency is higher (more than double the amount), especially if syngas are utilized in high-performance plants and/or a combined cycle (after treatments to eliminate remaining traces such as dusts, heavy metals, tar), while gas emission is very low.

A New Paradigm of Sustainable Building Underway in Settimo Torinese, Under the Guidelines of Laguna Verde

Fig. 12.3 Settimo Torinese, Italy, new eco building under construction on world–famous writer Primo Levi's old industrial site

Fig. 12.4 Settimo Torinese, Italy, eco building under construction on industrial site

The project underway, designed within the guidelines of Laguna Verde pointed out above, concretely anticipates Laguna Verde and consists of the sustainable revitalization of the former chemical plant the world-famous writer Primo Levi worked in as a technical director. The target of the Municipality of Settimo was to transform this old industrial building into a modern, sustainable one: a new conception of passive building and a self-reliant energy machine.

The main feature of the project is the realization of a new, transparent, and living green shell, external to the existing building. Its functions are as follows:

– Greenhouse (east and west facades)
– Vertical winter garden (north)
– Active shading (south) for the production of hot water
– Active shading shingle on the roof

The bioclimatic greenhouse guarantees an eco-efficiency comparable to a passive system: in winter, with an external temperature of –8°C, in the greenhouse it is around 32°C; in summer, with an external temperature of 32°C, in the greenhouse it is 27°C.

The carrying frame is a skeleton of plywood with wall plugs made of big, high-efficiency, low-emitting glass panels. Indoor, the vertical green is visible and tangible (a particular variety of jasmine), creeping on vertical cables. The greenery is designed to burn off the carbon produced inside and to transform into oxygen. The existing wall is refurbished by means of cheap rice chaff blown inside, as well as other natural materials. The flat roof is turned into a green roof, endowed with a system of rainwater catch and shaded by photovoltaic shading systems. The south side of the shell is shaded by high-performing vacuum solar panels, aimed at partially satisfying the demand of hot water (both for sanitary use and for the underfloor radiant panels). Most part of heating and cooling needs is satisfied by two heat pumps with two water probes and two geoprobes on a double layer under the stretch of water, whose function is also that of a heat dissipater. There is a double water network distribution, and the whole amount of water needed for toilets, cleaning, and watering is recovered from within the building. The air vent system is mechanized. The electrical energy needed for the building is produced out of a semi-transparent photovoltaic system on the roof (about 100 KWp).

The result is an active/passive "agile energy system," sustainable both on environmental issues (the building upgrades from class G to class A), and as far as economic issue is concerned, considering that there is a saving as high as 30–40%, on energy production systems (invested on the shell), as well as a proportional reduction of maintenance and no energy bill at all.

Conclusion

Sustainability is an objective to be attained, at all costs: not an easy target, but a feasible, realistic one. A parallel, much bigger twin project with Settimo's Laguna Verde is about to start in China, arising from similar premises and guidelines as Laguna Verde's, though on a completely different scale: ecosustainability and dialogue in order to build a new, innovative, and livable city of the future, based on its own natural resources.

The project is CAOFEIDIAN,[4] located 80 km south of Tangshan. A great endeavor, a major challenge, and a most important step toward the concrete realization of global sustainability in a big city: from Settimo Torinese's Laguna Verde, a small eco-town, to Tangshan's CAOFEIDIAN, a big Chinese eco-city.

What can be a better opportunity than a new, greenfield city like CAOFEIDIAN to conceive the most innovative, 100% ecological, renewable city, starting from its energy needs and the resources of the territory?

It is a new city of the future, whose challenge (and opportunity) is to derive its energy needs from what the territory can give: wind, sun, tidal movements, sea waves and streams, geothermal, and wastes.

Chapter 13
Sustainable Development in Lithuania

Natalija Lepkova

In 1992, the Baltic States, together with 175 other states, adopted the United Nation's plan *Agenda 21* that sets measures to be taken by the UN, governments, and other executive authorities to achieve a sustainable development of society and economy and to ensure the optimal use of the environmental resources.

N. Lepkova (✉)
Vilnius Gediminas Technical University, Vilnius, Lithuania, LT
e-mail: natalija.lepkova@st.vgtu.lt

W.W. Clark II (ed.), *Sustainable Communities*, DOI 10.1007/978-1-4419-0219-1_13,   193
© Springer Science+Business Media, LLC 2010

In 2003, the Republic of Lithuania approved a national strategy for sustainable development. One of the main priorities is to ensure the connection between social development and maintaining the environment and promoting the economy.

National Strategy for Sustainable Development

The National Sustainable Development Strategy identifies sustainable development as a compromise between the environmental, economic, and social objectives of society. These objectives would ideally provide an opportunity to improve the welfare of present and future generations without exceeding allowable limits of the impact to environment.

Preparation of the National Sustainable Development Strategy was based on the approach that Lithuania's development during the investigated period (until 2020) will be mainly influenced by eurointegration processes. Thus, foreseen changes and their implementation directly correlate with the Lithuanian membership in European Union.

The strategy takes into account peculiarities of Lithuania, as a country with an economy in transition. While assessing the changes in consumption and environmental pollution at the present developmental stage of Lithuania, the main focus is an "eco-effectiveness indicator" – consumption of energy and other natural resources per GDP unit, emission of pollutants and greenhouse gas per GDP unit, per conditional energy consumption unit, and so on.

The strategy foresees significant reductions in the consumption of energy and other natural resources as well as in emissions of pollutants and greenhouse gas emissions per GDP unit. *The main objective of sustainable development in Lithuania is to achieve the present developmental level of EU countries by 2020, according to indicators of economic and social development as well as efficiency in consumption of resources.* In order to achieve these objectives, it is necessary to base future development of the economy on advanced, environmentally friendly technology. Therefore, the strategy will put special emphasis on the design and implementation of technologies that are based on scientific achievements and knowledge rather than resource-intensive technologies [1].

Special efforts have been made to link the objectives and tasks of different sectors due to the cross-sectoral barriers within sustainable development. For example, if it is foreseen that expansion of the amount of rapeseed and crops cultivated for energy purposes (biodiesel and bioethanol) would equate to 15% of transportation fuel needs, then the industrial sector would have to promote the strengthening of required capacities for this expansion of biofuel production. Furthermore, the transportation sector must expand a network of petrol stations that sell this type of fuel, and the financial sector must establish economic preconditions for the increased biofuel competitive ability. Similarly, if the waste management sector expects to increase recycling capacities for secondary raw materials, including packages, the industrial sector must promote the development of enterprises that use local secondary raw materials, and so on. In addition to commercial sectors, active public involvement within these processes is anticipated. Public involvement includes sorting of household waste, purchase of products made from secondary raw materials, use of biofuel, and the encouragement of environmentally friendly lifestyles.

The implementation of sustainable development principles is impossible without extensive public participation. The strategy foresees the encouragement of active public participation not only in addressing specific above-mentioned tasks, but also in decision-making from different levels ranging from the National Commission for the implementation of the Sustainable Development Strategy to local authorities [1].

Lithuanian Sustainable Development Priorities and Principles

Lithuanian sustainable development priorities and principles were formulated by assessing the interests and peculiarities of the country as well as provisions of the EU Sustainable Development Strategy and other documents.

The EU Sustainable Development Strategy is confined to a limited number of problems and identifies six sustainable development priorities including mitigation of global climate change, minimization of the impact of the transportation sector on the environment, minimization of threats to public health, more effective use of natural resources, combating poverty and social exclusion, and addressing problems of an aging society.

If compared with other EU countries, the range of development problems is much wider in the country's transition economies. Because Lithuania is considered to have a transition economy, it has expanded its list of sustainable development priorities. First of all, rapid and stable economic growth is an essential precondition for a successful sustainable development provisions.

Within the long-term economic development strategy of Lithuania, the most acceptable economic development scenario in terms of sustainability concerns the anticipation of an annual GDP growth of 5–6%. A resulting growth through the sustainable development strategy period (until 2020) would make it possible for Lithuania to reach the present average level of economic development required for EU countries. The slow economic growth (pessimistic) scenario would not ensure implementation of the main sustainable development strategy objective that is formulated in the introductory chapter of the strategy. According to the forecast, the rapid economic growth (optimistic) scenario would pose a threat of a rapid increase in environmental pollution. Therefore, moderate economic development balanced between economic sectors and regions is considered as one of the most desirable sustainable development priority of Lithuania [1].

The household sector is often neglected when talking about the impact from different economic branches to the environment. Given that the Lithuanian household sector consumes more than one-third of the final energy production and it is considered to be one of the major water polluters, the national sustainable development strategy emphasizes the integration of environmental concerns into the household sector as well as minimization of impact from this sector to the environment [1].

Today in Lithuania, public awareness of the main sustainable development principles is insufficient. Thus, public education including environmental education and the promotion of environmentally friendly lifestyles are one of the primary tasks of sustainable development. Enhancement of the scientific research role, more effective application of research results at practical levels as well as design and implementation of environment-friendly production, and information technologies are all considered to be important priorities [1].

Sustainable Development Vision and the State Mission

The vision provides a general picture of Lithuania's future with a special emphasis on forthcoming changes and features that reflect the process and results of development. Although the vision is not presented in a certain time frame, formulation of the general vision and the vision of the main sustainable development components (environmental protection, economic and social development) is mostly based on the implementation period until 2020 [1].

Success of the sustainable development strategy can be ensured only if its main provisions and ideas are understood and supported by society as a whole. Nevertheless, the role of state institutions is very important. The state mission in sustainable development outlines the main objectives of the state in implementation of the strategy. The vision and mission are closely connected by their structure and approaches. However, the state general objectives and actions are more specifically formulated in the mission that helps to implement the main provisions of the vision [1].

Sustainable Development Vision

According to social and population health indicators, Lithuania will reach the current average level of the European Union countries. One will see an increase in the birth rate, and population will increase along with improving welfare. Youth emigration will decline, and rather large part of emigrated young people will return to Lithuania (see Table 13.2). With ongoing policy for promotion of the development of problematic territories, imbalanced regions will significantly decrease and high level of employment and living standard as well as good health care will be guaranteed. All strategic decisions related to economic activities and territories planning will be taken based on the results of environmental and health impact assessment. Preconditions for mitigation of health inequity, improvement of health care accessibility, and quality of services will be established. Improved public information and participation will increase public awareness in protecting and strengthening of their health. A modern education system will guarantee such education, which corresponds to the EU level. It will form a dynamic and responsible society, which actively participates in very important decisions at municipal and state levels. Improved working conditions, living standard, and health protection will influence prolonged age up to the average in EU. Proper pension security will be secured by combining current and accumulative financing. Many people will have a guaranteed possibility to work and earn, and social support will be provided only to those who really need it. Strong emphasis will be paid to disability, poverty, and social exclusion prevention; extreme poverty will be eradicated. A rational system for housing, its improvement, and energy efficiency increase will be established and available to everyone. Strong consideration will be given to preservation of Lithuanian ethnic regions, cultural identity of national minorities.

Protection and management of tangible cultural values will receive sufficient support from the state. This will help to preserve and revive valuable historical heritage for future generations. Protection and management of cultural values will become an integral part of territorial planning inside the country [1].

Active state regional policy in the field of regional sustainable development will promote more balanced development of Lithuanian regions. Strengthening of the development of problematic territories will significantly reduce regional imbalances, guarantee high employment rate and living standards as well as good health protection to all people, and ensure a healthy environment. Rational cooperation between the state, regional institutions, municipalities, and society will base on the partnership principle [1].

The State Mission

In order to execute the sustainable development strategy, the state must draw on different legal, economic, and organizational measures, as well as state and public institutions. Some European countries have already made constitutional amendments or adopted constitutional laws that enforce the main sustainable development principles at the highest juridical level. In order to ensure stability and accessibility of the state policy from a sustainable development point of view, the main provisions of Lithuania should be not only legitimized in this strategy but also integrated into special sectoral plans, programs, regional and municipal planning documents, and other relevant legal acts.

The mission of state in the field of environmental protection is to control and regulate the impact on environment and direct economic subjects and state institutions to the prevention of negative impacts on the environment and human health instead of putting major efforts into the elimination of negative impact consequences. Consistent implementation of EU directives and national laws regulating the impact on environment and human health and increase in producer responsibility for environmental pollution are a few tasks. While preparing for decommissioning of Ignalina NP, institutions responsible for economic development and environmental protection must initiate and ensure timely preparedness of thermo-electrical power plants to supply Lithuania with electricity with minimum increase in pollutant emissions. It is very important to effectively allocate state financing and EU assistance for modernization of the water supply and sewerage infrastructure, the establishment of water resource management system based on the river basin principle, as well as establishment of modern waste management system. Preparation of the program for growth of forest area in Lithuania and its implementation will not only facilitate the effective use of nonproductive agricultural land but also add to missing elements of the Nature Frame, support integration of Lithuanian protected areas system into the European ecological networks by establishing required connections [1].

The state must ensure protection of natural resources and effective and more economical use of these resources by legal and economic measures. Pollution charges have to be restructured so that their size would depend not only on consumed amount of resources but also on consumption efficiency. The state must support the broadest use of local renewable resources and recycled materials in every possible way. Municipalities must be given the better right to dispose of natural resources within their territories. Local renewable biological resources must meet up to 15% of Lithuanian energy demand. A strong legal base must be prepared and every kind of support allocated to develop agriculture for energy purposes and establish a recycling industry. This will provide possibilities to meet a similar part of fuel demand for the transportation sector by producing biodiesel and bioethanol. As demonstrated by the experiences of other countries, the main impediment for broader use of alternative energy sources (biofuel, wind, hydroenergy, etc.) is resistance by companies producing and selling oil products. The state must urgently regulate legal and economic aspects of alternative energy production and utilization.

The main state mission in the economic sector from a sustainable point of view is to promote well-balanced development of different economic branches and reduce interdepartmental barriers by legal and economic means. The optimal distribution of state budget allocations and EU assistance for sustainable development provisions is one of the major tasks of the state. Investments and economic support must be directed not only toward the increase in economic production efficiency and quantitative growth but also toward an increase in ecological production efficiency and reduction in environmental degradation and human health [1].

The main state mission in the social field is to provide each inhabitant of the country who is able and willing to work with a possibility to ensure a proper standard of life by their own income. The active employment policy must develop permanent access to all education systems that could ensure proper labor force qualifications to meet the needs in a changing labor market. The state social policy, while cooperating with social partners, should take into account all inhabitants in the country, ensure proper development of social security from the main social risk factors, promote motivation for economic activity, and provide social support only to those who need it the most. The main emphasis will focus on the prevention of poverty and social exclusion. The pension reform will provide possibilities for more effective coordination of current financing with accumulative financing and assure prosperous old age.

The state mission in the public health sector is to regulate and control environmental impacts on public health and to ensure integration of health-related concerns into decisions taken at all levels and in all fields of activities. The state should constantly encourage such activities, which help to protect and improve public health, prohibit and limit harmful activities, ensure compensation of damage to public health, and support implementation of prophylactic programs. Science and education must become one of the most important priorities of the development of state. All Lithuanian citizens must be secured with a possibility to attain higher education. Encouraging civic activity, promoting of environmental awareness education, and adapting health-friendly lifestyles must be emphasized on all education levels. Integration of science and scientific achievements into everyday life, formation of a knowledge-based society, and design and introduction of advanced environment-friendly technologies must become one of the important state priorities. In the light of faster eurointegration and globalization processes, a significant state mission encourages the preservation of Lithuania's cultural identity while maintaining continuity with national history [1].

The state must be concerned about its historical and cultural heritages by improving the legal framework and institutional systems of cultural protection and management, creating favorable conditions for investments of protection and use of cultural heritage.

The state mission in the field of sustainable development of regions is to ensure harmonized development of all regions in order to reduce disparities between economic development and individual welfare. The territorial administrative governance reform must continue regional and local institutional capacities to use possibilities and advantages of the EU membership in the most efficient way. The state must support the development of regional and local centers and strengthen the national system of territorial planning by legal, administrative, and economic means [1].

Strategy Implementation and Control

First of all, national programs have to be specified and amended with measures included in the strategy. The planned measures for regional sustainable development have to be included into regional development plans. Secondly, an effective monitoring system providing an opportunity to regularly assess achieved progress and identify obstacles and problems has to be created. Based on monitoring results, supplementary measures to ensure implementation of the strategic objectives can be planned. Considering ongoing internal and external changes, the strategy has to be regularly revised and amended.

A National Commission of Sustainable Development chaired by prime minister of the Republic of Lithuania was formed in 2000 in Lithuania to ensure coordination of sustainable development process at the highest level. The commission is composed of representatives of ministries, the president's office, and various other institutions and public organizations.

Special scientific institutions and scientific subdivisions are working in many countries to monitor the progress of the strategy and to perform a thorough analysis of ongoing changes and causalities. Based on performed analysis, they provide state institutions with suggestions and recommendations on improvement of the strategy, and they participate in preparation of sustainable development reports and other official documents. The establishment of such a scientific subdivision (a group of scientists) in Lithuania is also recommended.

Biennial sustainable development reports shall be discussed in the National Commission on Sustainable Development and submitted to the United Nations and EU institutions as well as presented to the public [1].

Sustainable Development Indicators

In order to effectively monitor the development of the strategy, simple quantitative sustainable development indicators have to be defined. These indicators must be directly linked with objectives and targets outlined in the strategy to provide a possibility for a regular assessment of achieved progress.

A list of national sustainable development indicators provided below takes into account indicators recommended in EU documents and the national specifics of Lithuania.

Today, different classifications of sustainable development indicators are proposed. In the National Sustainable Development Strategy they are grouped according to three main sustainable development sectors: environmental quality, economic development, and social development. Such a type of grouping is rather conditional as quite many of presented indicators are cross-sectoral describing an interaction between sectors. Regional development indicators are specified in a separate group. They reflect situation in smaller territorial administrative units [1].

Indicators of state of environment:

- *Emission of greenhouse gases in CO_2 equivalent:* total (million tons) and per area unit (km^2) as well as per GDP unit (in total and according to the branches of economic activities).
- *Emission of acidifying compounds (SO_2, NO_x), ground-level ozone precursors (NO_2, non-methane volatile organic compounds):* total (thousand tons) and per area unit (km^2) as well as per GDP unit (in total and according to economic activities).
- *Amount of discharged wastewater* (thousand tons): *a part of wastewater treated according to EU standards* (in percentages).
- *Area of eroded land:* thousand hectares and percentage from area of agricultural land.
- *Amount of household waste:* total (thousand tons), kilogram per capita, and ratio with average expenditures of consumption.
- *Amount of sorted household waste during the primary sorting:* thousand t. and percentage from total amount of household waste.
- *Amount of industrial waste:* total (thousand tons) and per GDP unit (in total and according to the branches of economic activities).
- *Amount of hazardous waste:* total (thousand tons) and per GDP unit (in total and according to branches of economic activities).
- *Public waste management service supply:* percentage from total number of population [1].

Indicators of economic development. The economic foundation of any society lies in its assets created by people's work or by production. As the economic growth is analyzed as a positive indicator by itself, to create economic foundation for the sustainable development, the Brundtland Commission recommended 5–6% growth for developing countries within the immediate 30 years, and 3–4% for industrialized countries. This growth should be based on the instillation of new ecological technologies, and its environmental impact should be subject to strict regulation.

The indicators of economic development are as follows [1]:

- *Gross domestic product at comparative prices;* million Lt (Litas – Lithuanian currency; 1 euro = 3,46 Litas, according to 2008 year official currency rate); *GDP increase*, in percentages (in total and according to branches of economic activities).
- *GDP per capita:* Lt.

The gross domestic product is shown in Table 13.1.

Table 13.1 Gross domestic product [2]

	At current prices, LTL mill	Compared to the previous year, growth, drop (−), %	Per capita, at current prices, LTL
1995	25956		7152
1996	32740	5,1	9090
1997	39998	8,5	11188
1998	44699	7,5	12594
1999	43667	−1,5	12391
2000	45674	4,1	13052
2001	48585	6,6	13956
2002	51971	6,9	14981
2003	56804	10,3	16445
2004	62587	7,3	18217
2005	71200	7,6	20854

Source: Statistics Lithuania (2008)

- *Amount of final energy consumed in production:* total (thousand tons by oil equivalent) and per GDP unit (in total and according to branches of economic activities).
- *Amount of water consumed in production:* total (thousand tons) and per GDP unit (in total and according to branches of economic activities).
- *Distribution of cargo and passenger transfers according to modes of transport:* thousand ton kilometers, thousand passenger kilometers, and in percentages.
- *Investments into the development of different modes of transport:* million tons and percentage according to modes of transport.
- *A part of biofuel in total amount of fuel consumed in transport sector:* thousand tons and percentage.

- *Density of motor vehicles:* number per 1000 inhabitants.
- *Density of cars:* number per 1,000 inhabitants.
- *Road network:* kilometer of roads per 100 km^2.
- *Railway network:* kilometer of railway per 100 km^2.
- *Number of enterprises where cleaner production methods are introduced:* units and a part of total number of enterprises in percentages.
- *Production based on advanced technologies:* percentage from GDP produced in industrial sector.
- *Part of renewable energy in total primary energy consumption:* percentage.
- *Part of electricity produced from renewable energy sources in total electricity consumption:* percentage.
- *Area of crops for production of biofuel;* 1,000 ha and percentage from total area of farming land.
- *Area of ecological farms:* 1,000 ha and percentage from total agricultural land.
- *Use of pesticides:* total amount according to active substance and kilogram per hectare.
- *Use of mineral fertilizers:* total amount according to active substance and kilogram per hectare (in total and according to substances [NPK]).
- *Amount of water consumed in housing:* total (thousand tons) and liters per inhabitant per day.
- *Amount of heat consumed in housing:* million tons by oil equivalent.
- *Part of household expenditures for utility charges:* percentage from average household income.

Indicators of social development [1]:

- *Economic activity and employment rate:* percentage from employable people.
- *Unemployment and long-term unemployment rate:* percentage from labor force (see Tables 13.2 and Table 13.3).
- *Rate of expenditures for social security:* percentage from GDP.
- *Poverty rate:* number of people in percentages living below the poverty level.
- *Indicator of inequality of people income:* income differentiation coefficient.
- *Average life expectancy:* years (in total and according to genders).
- *Growth of population:* percentage.
- *Mortality of newborns up to 1 year:* number per 1000 newborns in a given year.
- *Average living space per capita:* m^2.
- *Proportion of population living in housing of inadequate quality:* percentage.
- *Mortality from accidents at work:* number of people killed per year.
- *Allocations to education and science:* part of GDP in percentages.
- *Allocations to culture:* part of GDP in percentages.
- *Investments into science and technology development;* million Lt.
- *Number of students:* total number of graduated students in a given year and their part from young people of that age in percentages.
- *Number of pupils at secondary schools:* total number of pupils graduated in a given year and their part from people of that age in percentages.

Table 13.2 Main indicators of economic and social development (annual) in Lithuania [3]

Indicators	2003	2004	2005	2006	2007
Annual average population number (thousands)	3454,2	3435,6	3414,3	3394,1	3366,2
Unemployment rate, by labor force survey data, %	12,4	11,4	8,3	5,9	4,3
Inflation (December compared to December of previous year) (%)	−1,3	2,9	3,0	4,5	7,3
Average monthly gross earnings of employees in the whole economy (LTL)	1072,6	1149,3	1276,2	1500,2	2052,0
Gross domestic product at current prices (LTL million)	56804	62587	71200	81973,6	77939

Source: Statistics Lithuania (2007)

Table 13.3 Employment of Lithuania by type of activity [4, 5]

Type of activity	2003	2004	2005	2006	2007
Total workforce (thousand)	2319.9	1620.6	1606.8	1588.3	1534.2
Agriculture, hunting, forestry, and fishing (%)	17.9	15.8	14.0	12.4	10.4
Industry and construction sectors (%)	28.1	28.2	29.1	29.7	30.7
Service and commercial sectors (%)	54.0	56.0	56.9	57.9	58.9

Source: Statistics Lithuania (2007, 2008)

- *Number of pupils at vocational schools:* total number of pupils graduated in a given year and their part from people of that age in percentages.

Indicators of regional development [1]:

- *GDP of a region per capita and its ratio with the national average:* percentage.
- *Foreign investment:* million Lt. and a part from total foreign investments in the country in percentages.
- *Part of employable population:* percentage and ratio with average of the country.
- *Unemployment rate:* percentage and ratio with average of the country.
- *Part of people living below poverty level:* percentage and ratio with average of the country.

- *Emission of pollutants in the region:* t./km^2 and ratio with the national indicator.
- *Forestry area in the region:* percentage.
- *Number of municipalities where process of Local Agenda 21 has been initiated:* units and a part from total number of municipalities in percentages.

The main economic and social indicators for Lithuania are shown in Table 13.2.

According to Table 13.2, the average population and unemployment rate in recent years has decreased. The main reason is high emigration levels, especially after joining the EU in 2004 (May 1). In 2007 year the inflation rate in Lithuania was very high (about 7,3%), creating a barrier for Lithuania to introduce euro. Gross domestic product has increased and influenced the average monthly gross earnings of employees. The population employment of Lithuania by type of activity is shown in Table 13.3

From Table 13.3 it can be seen that the number of total workforce decreased. But distribution of the workforce by sectors is different. The higher increase is in industry and construction sectors, corresponding to the main tendencies in EU economy.

Sustainable Development Models

The goal of sustainable development is to combine economic growth, social progress, and the sparing use of natural resources while maintaining an ecological balance and ensuring favorable living conditions for current and future generations. Development is fostered in its natural environment; thus, it is important to find out reasonable extent and form of development, so that quality of life is maintained and the negative impact on environment is reduced.

Sustainable development is a way to combine two different and sometimes contradictory approaches: "development-progress-growth" and "stability-safety-environment." Brundtland Commission created this dilemma and was the first to formulate the goal of the sustainable development. In Fig. 13.1 the map of Lithuania is shown.

Analysis of Lithuania's sustainable development: urban development. Assessment of the development of Lithuania's towns and regions with the aid of social and economic data given by Statistics Lithuania revealed that development of individual towns and regions of Lithuania is very uneven and the standard of living is very different. Generalizing information drawn from 35 social and economic indicators, it turned out that economic growth in recent years concentrated in individual towns that predetermined a perspective for their further development [7].

Results given in Fig. 13.2 show that unevenness of development among Lithuania's towns was increasing in 1999–2001. Greatest changes were seen in socioeconomic situations and standard of living in Vilnius, Kaunas, and Klaipeda. Changes that started during the transformation from a planned economy to relations of market economy are still going on; population with better education migrate

Fig. 13.1   Map of Lithuania (*Source*: http://atlas.mapquest.com/atlas/?region=lithunia [6])

to big towns and surrounding regions. Development was and is very intensive in Vilnius. Improvement in the standard of living in Vilnius attracts immigration from other regions to work and live in Vilnius, resulting to a boost in housing prices in 2005.

Results of urban development given in Fig. 13.2 do not reflect such an important sphere for Lithuania as tourism and the related active cultural life are most active in resorts of Lithuania.

According to Fig. 13.3, tourism and cultural life development was very successful in resorts such as Palanga, Neringa, Birstonas, and Druskininkai. Intensive development of these resorts that started in 2001 still continues, which is evidenced by rapidly growing housing prices [7].

Analysis of evenness of Lithuania's development: foreign investment. Even greater differences in the standard of living are seen among towns and rural areas. Analyzing foreign direct investment in 1996–2001, it was noticed that in 1998, the possibilities for regions of Lithuania to attract foreign investment continued to decrease and in 2001 only two clusters remained (Fig. 13.4): the first cluster, covering Vilnius, Mazeikiai, Klaipeda, and Ignalina, features high investment levels, while the second cluster covers the remaining towns and regions with very low level of foreign investment.

Analysis of the standard of living and housing: Provision and quality of housing define the economic level reached by the country and the degree to which social needs of population are satisfied. Improvement in living conditions is related to changing esthetic and daily needs, as well as a supply of new domestic appliances.

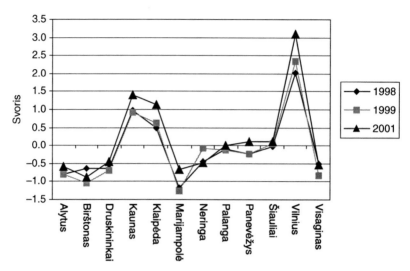

Fig. 13.2 Assessment of socioeconomic development and standard of living in 1998, 1999, and 2001 applying the method of a factorial analysis [7]

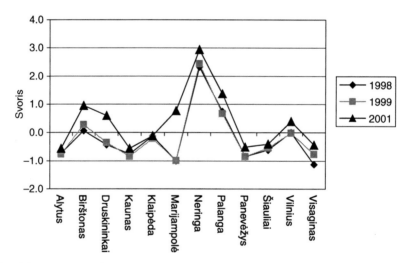

Fig. 13.3 Assessment of tourism development and cultural life in towns of Lithuania in 1998, 1999, and 2001 [7]

The changing standard of living is accompanied by a changing attitude toward housing requirements. Living conditions influence the demographic situation as well as family structures and relations. The idea that a decrease in the number of constructions reveals a deepening economic crisis proves to be true.

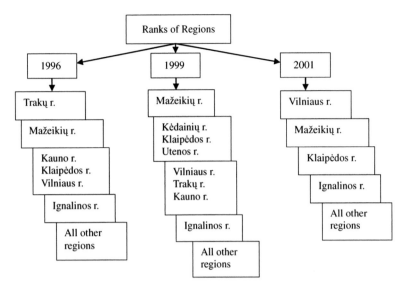

Fig. 13.4 Degrees of alternatives (regions) arranged in descending order according to factor coefficients [7]

Since 1990, the annual construction volume has been decreasing. At the same time, the contribution of housing constructions to the national economy and its strengthening has decreased. This influenced the deepening of socioeconomic problems. Such problems range over the lack of accessible housing for new households, limited mobility of labor, low level of economy and employment growth, ineffective energy consumption, and so on. Reasons for these problems are complex and may not be attributed to any single field.

The number of houses built in 1998 (4,176) is 5.5 times lesser than that in 1990 (22,100). The greatest decrease in the volume of housing construction was observed in 2001 (Fig. 13.5). In 2002, the number of dwelling houses started growing slightly. The annual construction of new houses accounts for 0.3%, and the annual turnover reaches 2.7% of the existing housing stock, while the EU average of this indicators is respectively 1.5 and 3.5% [7].

Figure 13.5 shows that construction volume has stabilized since 1995. The link between these two indicators is statistically significant only in the period 1990–1994. Since 1995, the statistically significant link has been observed between built housing and GDP [7]. The more detailed provision of housing in Lithuania is shown in Tables 13.4 and 13.5.

Emigration of population from regions to major towns increased the lack of apartments in several major towns of Lithuania: Vilnius, Kaunas, and Klaipeda, where most apartments were being built. In 2002–2005, small volume of constructions and growing GDP caused a dramatic price increase [7].

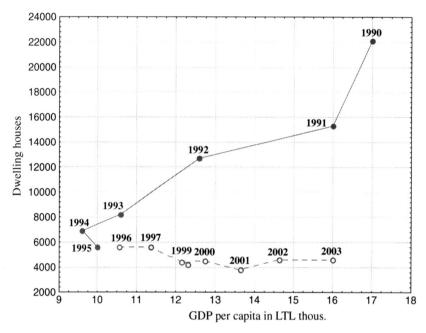

Fig. 13.5 Change in housing built and GDP at constant prices 2,000 per capita in LTL thousand in 1990–2003 (1 USA dollar = 2.34 LTL (Lithuanian Litas) [7]

Table 13.4 Dwellings built in 1995–2006 [3]

	1995	1996	1997	1998	1999	2000	2001	2002	2003	2004	2005	2006*
Dwellings	3368	3281	3173	2160	2580	2559	1987	2563	2535	3920	3250	4512
Per 1,000 inhabitants	1.5	1.6	1.6	1.2	1.2	1.3	1.1	1.3	1.3	2	1.7	2.1
Useful floor space of dwellings (1,000 m^2)	207	209	210	167	225.9	185.5	129.5	182.3	179.7	256.4	651.6	770.8
Average floor space per dwelling (m^2)	62	64	66	77	87.6	72.5	65.2	71.1	70.9	65.4	72.2	68.5

*Unofficial source: Statistics Lithuania (2007)

Table 13.5 Single-family houses built in 1995–2006 [3]

	1995	1996	1997	1998	1999	2000	2001	2002	2003	2004	2005	2006*
Houses	2232	2343	2389	1890	1674	1904	1798	1999	2093	2884	2683	7292
Useful floor space of houses (1,000 m^2)	358	422	397	325	292.3	321.1	255.8	277.8	311.7	442.7	416.8	462
Average floor space per house (m^2)	160	180	166	172	174.6	168.6	142.3	139		148.9	153.5	155.4	166.2

*Unofficial source: Statistics Lithuania (2007)

Sustainable Development Policy in Lithuania

The Lithuanian Parliament approved the Environmental Protection Strategy of Lithuania in 1996 after the reestablishment of the independence of Lithuania on the basis of a new national economic development policy.

The government adopted the "Action Programme" aimed at directing the country toward sustainable development so that clean and healthy environment and biological and landscape diversity could be preserved and effective consumption of natural resources ensured. The environmental protection strategy was based on the principles laid down in the Rio de Janeiro Declaration. Later, based on the principles of sustainable development, strategies for certain spheres were mapped out, for example the National Strategy for the Implementation of the United Nations Framework Convention on Climate Change, the Biological Diversity Conservation Strategy, the Strategy for the Public Environmental Education, the Energy Strategy, and so on [8].

International Cooperation

Until 1990 Lithuania had no practical possibility to establish closer cooperation with foreign countries. The information about the state of the environment in the former Soviet Union was considered to be a state secret. However, even then with the help of bilateral agreements on environmental protection among the former Soviet Union, the USA, and Sweden, relations between Lithuanian and foreign environmental protection specialists were established. Their experience and ideas were taken over and joint experiments and measurements of environmental pollution commenced.

The situation changed drastically in 1990 after the reestablishment of independence. With the assistance of foreign experts, new laws and legal acts were created. Lithuania's path to sustainable development started in 1992, when the Program for Environmental Protection of Lithuania was devised. It defined basic environmental problems within the country. This program was based on sustainable consumption of natural resources and on the condition that any economic activity should comply with environmental requirements.

In 1992 Lithuania joined the family of the Baltic Sea Region countries by signing a multilateral agreement – the Convention on the Protection of the Marine Environment of the Baltic Sea Area (Helsinki). In the same year, together with other eight Baltic Sea Region countries, Lithuania started implementing the Joint Environmental Action Program of the Baltic Sea. All the Baltic Sea Region countries actively cooperate in carrying out the program as well as the Helsinki Convention and all its annexes. According to a gradually changing situation, the Helsinki Commission improves different execution plans of the convention and adopted recommendations by harmonizing them with the sustainable development principles [8].

For example, alongside the permanently emphasized chemical pollution (nutrients, heavy metals, and persistent organic pollutants), a negative environmental impact is becoming a threat (especially in the coastal zone) due to urban development. With the point pollution decreasing, nonpoint pollution resulting from agricultural activity and development of the transportation sector poses one of the most serious problems. Problems of production and consumption are of ever-increasing concern, especially due to unregulated waste management. There is a pressing concern to curb the issues related to fishing in the Baltic Sea as well as to deal with pollution from ships. On the basis of the principles of sustainable development, Lithuania and other Baltic Sea Region countries need to address these and similar issues [8].

In order to strengthen relations with other European states, Lithuania signed two conventions of the United Nations Economic Commission for Europe – the convention on the Protection and Use of Transboundary Watercourses and International Lakes and the convention on the Transboundary Effects of Industrial Accidents at the beginning of 1992.

In 1992 the Lithuanian delegation took an active part in discussion and adoption of the main conference documents at the United Nations Environment and Development Conference held in Rio de Janeiro. It signed two new international conventions – Framework Convention on Climate Change and Convention on Biological Diversity. Prior to the Rio Conference, the National Report of Lithuania intended for the conference was prepared. It presented economic infrastructure (trade, industry, agriculture, energy), the state of environment (air, water, waste, landscape and land use, soil, natural resources and their protection, environmental pollution, and public health issues), environmental policy objectives and strategies, and implementation of environmental policy in different spheres.

Following the Rio de Janeiro Conference, Lithuania established permanent relations with the United Nations Sustainable Development Commission in New York and presented reports at several sessions of this commission.

During the meeting of the ministers of the Baltic Sea Region countries in May 1996, the initiative was launched to gradually implement sustainable development in the region by mutual efforts. The joint program for the region, Baltic 21, was completed within 2 years, and in June 1998 the council of the Baltic region countries (Ministers of Foreign Affairs) approved it. In essence, this is a sustainable development program for our region. Eleven countries – nine Baltic Sea Area countries, Norway, and Island – are all participating. The European Commission, several intergovernmental organizations, international financial institutions, and nongovernmental organizations are also members of the program. Lithuanian experts took an active part in drawing up this program and later participated in coordination. The program was drawn up for seven of the most important sectors of the Baltic Sea Region – energy, forestry, industry, transportation, tourism, agriculture and fishery, and urban planning. In 2000, the education sector was also included in the program. Each sector was allocated two countries responsible for the coordination of activities and implementation. Lithuania with Finland was in charge of the forestry sector, and Lithuania with Sweden was responsible for the educational sector.

Apart from the activities in separate sectors, Program Baltic 21 provides for joint actions uniting several sectors. Each sector is fully responsible for carrying out each sustainable development program.

In the forestry sector supervised by Lithuania and Finland, activities are carried out in accordance with sustainable forestry objectives: forests and forest lands must be managed and used in such a way and at such a pace that biological diversity, productivity, regeneration capacity, and vitality should all be preserved. Their abilities to carry out ecological, economic, and social functions at local, national, and global levels should be ensured so that they do no harm to other ecosystems. Sector plans are devised on the basis of the resolutions of the Rio de Janeiro, the Strasbourg (1990), the Helsinki (1993), and the Lisbon (1998) conferences on forest protection in Europe. Criteria and quantity indicators proposed by the European Union for sustainable development process of European forests are used in the sector [8].

In 2000 after the ministers of education signed the Hague Declaration, the Educational Agenda 21 (Baltic 21E) for the Baltic Sea Region was created. It was approved by all the ministers of education of the Baltic Sea Region countries at their second meeting in Stockholm. Lithuanian and Swedish specialists on education supervise that sector [8].

Permanent international cooperation of Lithuanian environmental specialists is not limited to the Baltic Sea Region. Such cooperation is promoted by designing plans for bilateral and interdepartmental agreements between Lithuania and other foreign countries. Urgent issues, such as transportation, environment and health, water and health, transportation and environment, energy and environment, agriculture and environment, and so on, are often addressed at the aforementioned events [8].

Main Instruments of Sustainable Development Policy in Lithuania

First and foremost, the main law – the Constitution of the Republic of Lithuania – regulates the rights and duties of individuals. The constitution contains several articles directly influencing the implementation of the principles of sustainable development in the country. Article 46 reads as follows: "Lithuania's economy shall be based on the right to private ownership, freedom of individual economic activity, and initiative. The State shall support economic efforts and initiative useful to the community. The State shall regulate economic activity so that it served common welfare of the people. The law shall prohibit monopolization of production and the market and shall protect freedom of fair competition. The State shall defend the interests of consumers." Article 48 states that "every person may freely choose an occupation or business and shall have the right to adequate, safe and healthy working conditions, adequate compensation for work, and social security in the case of unemployment." The constitution obliges the state to take care of the protection of natural environment. It also obliges the citizens to protect the environment. Article 53 establishes that "the State and each individual must protect the environment from harmful influences." Article 54 of the constitution establishes how the state must do that: "The State shall care about the protection of natural environment, its fauna and flora, separate objects of nature and particularly valuable districts; it shall supervise moderate use of natural resources as well as their restoration and augmentation. The exhaustion of land and earth entrails, pollution of waters and air, making radioactive impact as well as impoverishment of fauna and flora shall be prohibited by law" [8].

The law concerning environmental protection of the republic of Lithuania was adopted in 1992 and supplemented several times later and lays down the basic principles of environmental protection. The main objective of this law is to seek ecologically safe and healthy environments, to protect the characteristic landscape of Lithuania as well as diversity of biological systems. The law provides for environmental impact assessment, lays down the principle "polluter pays," and encourages the citizens and public organizations to ensure environmental protection [8].

On the basis of the Law on Environmental Protection, a number of other laws and legal acts regulating the consumption of natural resources and environmental protection were adopted. The basic laws related to consumption of natural resources and protection of the environment are presented in Table 13.6 [8].

Legal and natural entities, who have violated laws and other legal acts or have done harm to the environment, may be brought to criminal, administrative, or civil responsibility.

When integrating environmental requirements for the economic activity, the following economic instruments shall be applied:

- taxes on state natural resources;
- air, water, and soil pollution taxes;
- fines for exceeding the established pollution limits or concealed pollution, harm done to the environment, tree felling without a permit, an so on;
- excise/duties on fuel and cars;
- consumer taxes on water consumption, wastewater treatment, and household waste management;
- subsidies on project studies, research, and implementation of projects [8].

Environmental taxes on pollution and consumption of natural resources have been imposed since 1991. Taxes are collected to compensate for harm done and to cover the costs of restoration. On the basis of the amendments to the law of taxation on Environmental Pollution (1991, new wording in 1999) approved in 2002, the procedure of calculation and payment of taxes on environmental pollution as well as on some kinds of products and product packaging is being prepared.

Every ministry and organization subordinated to it as well as other state and municipal institutions implement laws and other legal acts within the area of their competence; formulate state policy in a relevant sphere; prepare standards, system of permissions, and economic measures; and ensure participation of the public in

Table 13.6 Basic laws regulating environmental protection and consumption of natural resources and other laws of significance for implementing sustainable development [8]

Law on Taxes on State Natural Resources (1991, amendments in 1996, 2000)
Law on Pollution Tax (1991, new wording in 1999, amendments in 2000, 2002)
Law on the Principles of Transport Activities (1991, amendments in 1997)
Law on Environmental Protection (1992, amended and supplemented in 1996, 1997, 2000, 2001)
Law on Tax on Oil and Gas Resources (1992, amendments in 1996)
Law on Protected Areas (1993, new wording in 2001)
Law on Land (1994, amended and supplemented in 1995, 1996, 1997, 1999, 2000, 2001)
Law on Forestry (1994, new wording in 2001)
Law on Territorial Planning (1995, amended and supplemented in 1997, 2000, 2001)
Law on Plant Protection (1995, new wording in 1998, amendments in 2001)
Law on Earth Entrails (1995, new wording in 2001)
Law on Local Government (1995, new wording 2000)
Law on Energy (1995, amended and supplemented in 1996, 1997, 1998, 1999, 2000)
Law on Provision of Information to the Public (1996, new wording in 2000)
Law on Construction (1996, new wording in 2000)
Law on Environmental Impact Assessment of the Proposed Economic Activity (1996, new wording in 2000)
Law on Water (1997, amendments in 2000)
Law on Protection of Marine Environment (1997)
Law on Wildlife (1997, new wording in 2001)
Law on Environmental Monitoring (1997)
Law on Waste Management (1998, amended and supplemented in 2000, 2002)
Law on Public Administration (1999)
Law on Ambient Air Protection (1999)
Law on Radioactive Waste (1999)
Law on Regional Development (2000)
Law on the Right to Access of Information from State and Municipal Institutions (2000)
Law on Biofuel (2000)
Law on Genetically Modified Organisms (2001)
Law on the Management of Packaging and Packaging Waste (2001, enters into force in 2003)

the process of adopting decisions related to any environmental impact and social issues.

The functions of the regional departments of the Ministry of Environment comprise issuance of permissions, environmental impact assessment, laboratory control, and application of environmental standards. In order to carry out these functions the regional departments have central apparatus and regional environmental protection agencies. Inspectors have the right to check plants and equipment in them. They can authorize laboratories to observe pollution and can impose fines if standards or permissions are violated. Inspectors from the regional departments and agencies are responsible for exerting control over application of environmental legal acts, norms, and standards. The process of applying the laws is based on the system of permissions and local monitoring. Environmental inspectors regularly check the amounts of emission harmful to the environment as well as the accuracy of the reports of plant operators. Nonprovision of data (or erroneous reports) on environmental condition and emission of pollutants without permissions are violations subject to punishment.

Territorial planning is one of the most important instruments for formulating and implementing the sustainable development policy. This activity is regulated by the Law on Territorial Planning and realized by preparing territorial planning documents. The law established the right and duty of the administration of all governing levels to prepare territorial planning documents for the territory it manages. Such a democratic and decentralized system creates a possibility to make use of all the public initiatives seeking to achieve the objectives set at the International United Nations Forum in Rio de Janeiro.

An important step to implement sustainable development ideas in relation to territorial planning is the preparation of the Master Plan of the Territory of the Republic of Lithuania undertaken on the basis of the resolution of the government. The plan was started to draw up in 1996. In the year 2002 it was submitted to the government of the Republic of Lithuania for endorsement and then it will be submitted to the Seimas for approval. A significant feature of that document is intersectoral planning methods seeking sustainable territorial development.

One of the necessary measures and preconditions for any proposed economic activity to be implemented in compliance with the principles of sustainable development is a qualified environmental impact assessment of that activity. In Lithuania environmental impact assessment is carried out in accordance with the Law on Environmental Impact Assessment of Proposed Economic Activity (2000). The law regulates the process of environmental impact assessment and relations between its participants. According to this law, the public plays an important role in the process of environmental impact assessment [8].

References

1. National Strategy for Sustainable Development. Approved by Resolution No.1160 of September 11, 2003 of the Government of the Republic of Lithuania, Vilnius, 2003, Internet access: http://www.am.lt/VI/en/VI/files/0.901665001073997792.pdf [viewed 2008 09 18].
2. Statistics Lithuania (2008). Sustainable Development indicators. Department of statistics to the Government of the Republic of Lithuania, Internet access: <http://www.stat.gov.lt/uploads/docs/Darnus_vystymasis_2006_internetui.pdf>, [viewed 2008 10 25].
3. Statistics Lithuania (2007). Economic and Social Development in Lithuania. Vilnius, 179 p.
4. Statistics Lithuania (2007). Statistical yearbook of Lithuania, Department of statistics to the Government of the Republic of Lithuania, Vilnius. Internet access: <http://www.stat.gov.lt/>, [viewed 2008 09 10].
5. Statistics Lithuania (2008). The main indicators of Employment of Lithuania by type of activity. Internet access: <http://www.stat.gov.lt/lt/pages/view/?id=2407>, [viewed 2008 09 11].
6. Map of Lithuania. Internet access: <http://atlas.mapquest.com/atlas/?region=lithunia>, [viewed 2008 09 25].
7. Vitalija Rudzkiene, Marija Burinskiene. Information Models for Assessment and Management of Development Trends. Summary of analysis in English. Whole book titled: Pletros krypciu vertinimo ir valdymo informaciniai modeliai (in Lithuanian). Vilnius: Technika, 2007.
8. National report on Sustainable Development. From transition to sustainable development. The republic of Lithuania. Vilnius 2002. p. 147.

Chapter 14
Rizhao: China's Green Beacon for Sustainable Chinese Cities

Calvin Lee Kwan

C.L. Kwan (✉)
School of Public Health, University of California, Los Angeles, CA, USA
e-mail: clkwan@ucla.edu

W.W. Clark II (ed.), *Sustainable Communities*, DOI 10.1007/978-1-4419-0219-1_14,
215

By sustaining an average annual GDP growth rate of over 10% during the last two decades, China has rapidly developed its economy [10], emerging as a key and powerful player in today's globalized market. However, alongside the rise of its economy, the country has seen an equally rapid decline in its environmental quality. Although China's environmental degradation and pollution are widespread, it has been estimated that over 75% of China's pollution can be attributed solely to burning coal [13]. The rapid economic development has created a burgeoning demand for energy, resulting in the construction of over 2,000 coal-fired power plants littered throughout the country, with plans for an additional 500 more by 2010 [12, 6]. Every day, particulate matter produced from these power plants choke the air, leaving a permanent brown haze over many Chinese cities. China is currently the world's second largest emitter of greenhouse gases behind the United States. In June 2008, China assumed the unenviable title of being the world's largest emitter of carbon dioxide. The International Energy Agency [9] estimates that in 25 years, China will emit twice as much carbon dioxide as all the countries of the Organization for Economic Cooperation and Development (OECD) combined. Given that China is already home to 16 of the world's 20 most polluted cities [7], it is clear that the country faces an arduous uphill battle to improve its environmental outlook.

With increasing attention and awareness to global climate change, countries and organizations around the world have begun to emphasize sustainable living and green development. China is no exception. In 2001, China's 10th Five-Year Plan called for increased investment in renewable technology, in particular solar, geothermal, and wind [15, 16]. China's recognition of the need to pursue renewable energies reached a new height in 2005 when the government drafted the Renewable Energy Policy. The policy introduced incentive structures, subsidies, and funds to promote development of renewable energy technologies. The policy was enacted into law on January 1, 2006, emphasizing and confirming China's commitment to renewable energy. This recent proactive trend continued through the planning of the latest and current 11th Five-Year Plan, also implemented in 2006. In this plan, China established medium- and long-range renewable energy goals; by 2010, 10% of all energy will come from renewable sources, increasing to 16% by 2020 [4]. To achieve the stated targets, China has committed over $260 billion USD toward developing its renewable energy infrastructure.

Despite these actions and objectives, the Chinese government realizes and acknowledges its shortfalls and challenges in national environmental protection.

> Environmental situation is still grave in China though with some positive development. The environmental protection targets of the "10th Five-Year Plan" period had not been met with 27.8% increase of SO_2 emissions and 2.1% reduction of COD as compared with that of 2000, while the targets should be 10% reduction.

> It is expected that the population of China will grow at 4% during the "11th Five-Year Plan" period with accelerated urbanization and over 40% growth of GDP. The contradiction between socio-economic development and resources and environment constraint becomes increasingly evident. International environmental protection pressure will grow and environmental protection of China is facing graver challenges.
>
> – The 11th Five-Year Plan for the Environmental
> Protection of the People's Republic of China

The urgency for action is becoming more pronounced as hundreds of thousands of Chinese citizens migrate from the suburbs and countryside into towns and cities. According to the Chinese National Statistics Bureau, in 1978, there were only 13 cities in China with populations of 100,000 or more. By 2008, that figure increased to 665 and is expected to reach 1,000 by 2015. China's current energy model of using fossil fuels to meet its uncontrolled energy demands of its cities is both unsustainable and infeasible. Recognizing this, the Chinese government identified 113 key cities, listed in Fig. 14.1, for environmental protection. Each of these cities will have strict environmental and energy conservation programs implemented. This includes developing renewable energy, minimizing energy consumption, and preserving local ecologies.

Municipality: Beijing, Tianjin, Shanghai and Chongqing

Provincial capitals: Shijiazhuang, Taiyuan, Hohhot, Shenyang, Changchun, Harbin, Nanjing, Hangzhou, Hefei, Fuzhou, Nanchang, Jinan, Zhengzhou, Wuhan, Changsha, Guangzhou, Nanning, Haikou, Chengdu, Guiyang, Kunming, Lhasa, Xi'an, Lanzhou, Xining, Yinchuan and Urumqi.

Cities under separate plan of the State Council: Dalian, Qingdao, Ningbo, Xiamen and Shenzhen

Other cities: Qinhuangdao, Tangshan, Baoding, Handan, Changzhi, Linfen, Yangquan, Datong, Baotou, Chifeng, Anshan, Wushun, Benxi, Jinzhou, Jilin, Mudanjiang, Qiqihar, Daqing, Suzhou, Nantong, Lianyungang, Wuxi, Changzhou, Yangzhou, Xuzhou, Wenzhou, Jiaxing, Shaoxing, Taizhou, Huzhou, Ma'anshan, Wuhu, Quanzhou, Jiujiang, Yantai, Zibo, Tai'an, Weihai, Zaozhuang, Jining, Weifang, Rizhao, Luoyang, Anyang, Jiaozuo, Kaifeng, Pingdingshan, Jingzhou, Yichang, Yueyang, Xiangtan, Zhangjiajie, Zhuzhou, Changde, Zhenjiang, Zhuhai, Shantou, Foshan, Zhongshan, Shaoguan, Guilin, Beihai, Sanya, Liuzhou, Mianyang, Panzhihua, Luzhou, Yibin, Zunyi, Qujing, Xianyang, Yan'an, Baoji, Tongchuan, Jinchang, Shizuishan and Karamay.

Fig. 14.1 Key cities identified in the 11th Five-Year Plan for environmental protection [4]

China will beef up the policy guidance for conservation and efficient use of energy; make more efforts in energy saving administration according to law; accelerate the development, demonstration and extension of energy saving technologies; make full use of market-based energy saving mechanism in order to reduce the emissions of greenhouse gases (GHGs). China will vigorously develop renewable energy, actively promote the development of nuclear power plants, accelerate the development & utilization of coal field gas to optimize energy mix. China will make more efforts in the development of biogas projects in rural areas and recycling & reuse of the biogas from urban landfill facilities in order to control the growth of methane emissions. China will continue its efforts in the development of key ecological projects such as reforestation and the conservation of natural forest resources in order to raise forest coverage, carbon sink and enhance the adaptation capacity. China will enhance the monitoring and statistic analysis on the emissions of GHGs.

– The 11th Five-Year Plan for the Environmental
Protection of the People's Republic of China

Although China has developed and adopted clear-cut environmental goals, achieving them is a completely different matter. It is well known that laws, regulations, and targets are often issued by Beijing, although with little guidance on the implementation methods [14, 3]. These are often left to the discretion of provincial, municipal, and city administrations. Given the relatively recent focus on and development of environmental laws, regulations, and goals, these localized administrations may be inadequately equipped to successfully launch environmental and renewable energy programs. Despite these challenges, one city in China has emerged from the brown haze, acting as the shining green beacon for China's environmental and sustainability movement. The city of Rizhao has the necessary combination of innovative thinking and government leadership that places it far ahead of all 112 other cities in terms of Chinese environmental and sustainable awareness.

Rizhao, aptly meaning "sunshine," is a major Chinese seaport located in the Shandong Province of Northern China, shown in Fig. 14.2. As its name suggests, the location of Rizhao provides it with ample solar radiation, and it has traditionally been called the city that receives sunshine first in China. With a population of over 3 million, Rizhao is a bustling port city, rich in history and culture. However, what sets Rizhao apart from other Chinese seaports is its commitment toward using renewable energy resources, particularly solar energy, to generate power. Rizhao sets the bar in China for using renewable energy in a country dominated by fossil fuel-based energy resources. Through strategic planning and investing for over 15 years, 99% of households in the city utilize solar thermal water heaters [2]. Although this figure drops to 30% of households in the suburbs and villages of Rizhao, it is still a remarkable feat. By using solar thermal water heaters instead of electric water heaters, the city saves over half a megawatt of energy each year. Solar panels can be seen glistening on the rooftops and sides of buildings in the core district. Streetlights, park

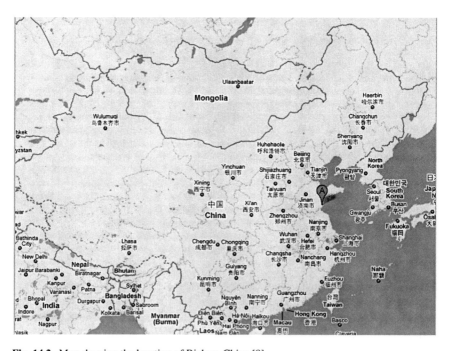

Fig. 14.2 Map showing the location of Rizhao, China [8]

Fig. 14.3 Rizhao rooftops with solar thermal water heaters [11]

Fig. 14.4 Closeup view of rooftop of solar thermal water heaters in Rizhao, China [5]

accent lights, and traffic signals utilize low-energy technology such as LED and are almost entirely powered by solar photovoltaic cells.

Rizhao's unique environmental progress was made possible by three key factors (Bai 2007):

(1) A government policy that encourages solar energy use and financially supports research and development
(2) Local solar panel and solar thermal heater industries that have developed and improved their products
(3) The political will of Rizhao city officials and leadership

These three factors have made Rizhao's transition toward a renewable energy infrastructure smooth and successful. The city's administration has been able to successfully incorporate and, more importantly, *carry out* renewable energy practices. Even the implementation model Rizhao utilized to develop its renewable energy program is noteworthy. Recently, governments such as those of the State of California, the United States, the UK, and Japan have provided subsidies and incentives to end users in order to promote the transition to renewable energy technologies, particularly solar PV. However, in the case of Rizhao, the city elected to place money into research and development at local solar PV and solar thermal heater manufacturers instead of for end-user subsidies. By doing so, the city was able to help manufacturers develop their technologies, improve upon them, guarantee their quality, and more importantly, lower the per-unit costs, making the technologies more affordable to all citizens. By bringing the price down to the same level as electric water heaters and coupled with the added incentive of additional annual savings through reduced energy consumption, it makes economic sense for citizens to purchase and install solar thermal water heaters.

To actively promote solar technologies and develop public acceptance, Rizhao regularly holds education fairs and exhibitions, encouraging solar manufacturers to show their products. City regulations also require that any new building or construction projects incorporate solar PV and solar thermal water heaters into their design. The construction process is heavily monitored by Rizhao officials to ensure proper installation and operation of the solar technology.

Interestingly enough, Rizhao does not rely solely on the sun for renewable energy. In fact, Rizhao officials have the belief and understanding that "waste" is simply resources or inputs that are in the wrong places. Using this philosophy, for the last 15 years, the government has been investing and developing a process to use "wastewater" from local businesses to generate marsh gas which can be used to produce energy. Currently, the city can produce over 4.5 million cubic meters of marsh gas annually, displacing over 36,000 tons of coal, making Rizhao a truly eco-city. However the quantity of energy produced is dependent on the purity of the marsh gas, which in turn is dependent on the quality of wastewater. Using another innovative approach, Rizhao offers tax rebates and credits to businesses for treating their wastewater before discharging. This helps to protect the local environment by removing contaminants from wastewater and has led to the development of new methods to produce and utilize marsh gas energy.

The combination of solar PV, solar thermal heaters, and marsh gas allows Rizhao to reduce its annual coal consumption by almost 1.5 million tons. This corresponds to annual carbon dioxide and sulfur dioxide reductions of 3.25 million tons and 21,300 tons, respectively. These results have led to Rizhao being consistently recognized as one of the top 10 cities in China in terms of air quality. In 2006, the State Environmental Protection Agency (SEPA) designated Rizhao as the Environmental Protection Model City of China. In addition to this, Rizhao has been recognized globally for its renewable energy program, receiving a 2007 World Clean Energy Award.

The case study of Rizhao shows that microscale or local renewable energy systems can play a significant role in reducing dependence on fossil fuels, improve the environment, and more importantly, *are reliable and feasible*. Although the technologies used in Rizhao will by no means replace the need for fossil fuel-based energy, it does have a measurable impact on the local environment. Even the simple installation of solar thermal water heaters has allowed the city to reduce its coal consumption by several thousand tons a year. Given that China is already the world's largest manufacturer of solar photovoltaic panels, solar thermal water heaters, and other renewable energy technologies, the country has an advantage over others in developing clean renewable energy cities like Rizhao. China needs to begin using Rizhao's renewable technology program as an example for future development of eco-cities. Several cities have taken notice of Rizhao, implementing similar plans. The city of Shenzhen in the Guangdong Province has mandated that all new construction buildings over 12 stories shall have solar thermal water heaters installed. In Shanghai, the local administration has made an effort to invest 1.5 billion dollars into solar PV for rooftops.

The successful development of Rizhao's renewable energy policies and programs should serve as an example for cities around the world. The fact that this city is located in China, arguably one of the most polluted countries in the world, gives hope that with the proper government structure, leadership, and innovative thinking, China can begin its campaign toward a low-carbon, renewable energy society.

References

1. Bai, X. (2007). "Rizhao: Solar, China: Solar-powered City." State of the World 2007: Our Urban Future 2007.
2. Bai, X., A. J. Wieczorek, et al. (2009). "Enabling sustainability transitions in Asia: The importance of vertical and horizontal linkages." Technological Forecasting and Social Change 76(2): 255–266.
3. Bardhan, P., D. Mookherjee (2005). Decentralization, Corruption and Government Accountability: An Overview. Handbook of Economic Corruption.
4. China (2006). The National 11th Five-year Plan for Environmental Protection (2006–2010), Ministry of Environmental Protection The People's Republic of China.
5. CSIRO. (2008). "Commonwealth Scientific and Industrial Research Organisation." from http://www.scienceimage.csiro.au.

6. CSPIN. (2008). "China State Power and Information Network." Retrieved October 2008, 2008, from http://www.sp-china.com/.
7. Economy, E. C. (2007). "The great leap backward? – The costs of China's environmental crisis." *Foreign Affairs* **86**(5): 38–59.
8. Google. (2009). "Google Maps." 2009, from http://maps.google.com.hk.
9. IEA (2009). "International Energy Agency."
10. NBRC. (2008). "China National Bureau of Statistics." Retrieved October 2008, 2008, from http://www.stats.gov.cn.
11. Sweeney, J. (2008). From Beijing to Lijiang. A. News, ABC News.
12. Watts, S. (2005). A Coal Dependent Future? BBC News. London, BBC.
13. Wu, Y. (2003). "Deregulation and growth in China's energy sector: a review of recent development." *Energy Policy* **31**(13): 1417–1425.
14. Young, A. (2000). "The Razor's Edge: Distortions and Incremental Reform in the People's Republic of China∗." *Quarterly Journal of Economics* **115**(4): 1091–1135.
15. Zhu, G. (2001). "Major Tasks of Environmental Protection of the Tenth Five Year Plan of China." China Council for International Cooperation on Environment and Development, from http://www.cciced.org/2008-02/03/content_9643326.htm.
16. Zhu, R. (2001). "Look into the Next Five Years (2001-2006)." Retrieved January 3, 2009, from http://www.china.org.cn/e-15/index.htm.

Chapter 15
Ecological Construction and Sustainable Development in China: The Case of Jiaxing Municipality

Wang Weiyi and Li Xing

Introduction

Jiaxing Municipality is located in the southeastern part of China's coastline and in the center of Yangtze River Delta. It abuts on Shanghai to the east, Suzhou to the north, Hangzhou to the west, and Hangzhou Bay to the south. It covers a land area of 3,915 km^2 and has a population of about 3.3 million.

W. Weiyi (✉)
Jiaxing University, Zhejiang, P.R, China
e-mail: anunique@yahoo.com.cn

W.W. Clark II (ed.), *Sustainable Communities*, DOI 10.1007/978-1-4419-0219-1_15,
© Springer Science+Business Media, LLC 2010

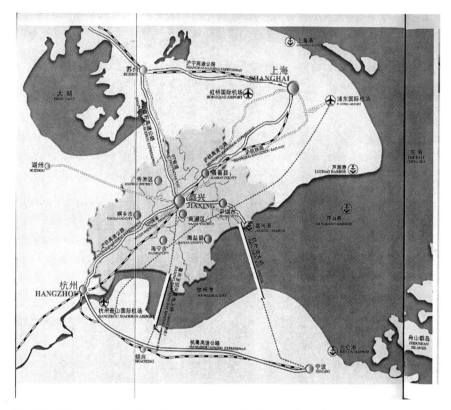

Fig. 15.1 The map on the left indicates one part of the Yangtze River Delta region. Jiaxing looks like a "middle kingdom" surrounded by China's fastest developing cities in the region

Jiaxing has a long history and culture. It is the cradle of culture in the south of the lower reaches of the Yangtze River. As early as 6,000 or 7,000 years ago, the ancestors bred the Majiabang Civilization, which is recognized as the representative of the culture of the Taihu Drainage Area in early Neolithic Age and as one of the origins of ancient civilization of Chinese nation.

Jiaxing has always been a rich and prosperous land and is historically known as China's "land of fish and rice" and "place of silk and satin." Since the founding of the People's Republic of China, especially since the implementation of the policy of reform and opening up to the outside world, jointly driven by the development of Pudong District of Shanghai, together with the opening up of the economy of southern Jiangsu Province as well as the booming of privately owned enterprises along the southern part of Zhejiang Province, Jiaxing has made a rapid progress in economy and its comprehensive strength has been enhanced remarkably. At present, all the counties and county-level cities under Jiaxing's jurisdiction are listed among the top 30 of the 100 wealthy counties in China and among well-off counties in

Zhejiang Province. In 2006, the whole city had a total output value of 158.51 billion yuan. In recent years, Jiaxing has actively promoted the harmonious development of economic, social, and ecological environments, and great achievements have been made in these aspects. Table 15.1 shows the economic development in 2008.

Table 15.1 Economic growth (unit: yuan RMB)

Item	Sum by the end of September 2008	Increase rate when compared with September 2007 (%)
GDP	128.7 billion yuan	11.7
Fixed assets investment	66.07 billion yuan	19
Local fiscal income	20.72 billion yuan	25.7
Per income for urban citizens	17,372 yuan	12.6
Per income for rural residents	8,189 yuan	14.1

Resource: Jiaxing Statistics Monthly Report, October 2008 [1]

Objective

The following sections provide empirical accounts of Jiaxing's achievements in sustainable development in terms of environmental protection and ecological embedding in many areas such as "new village project," "ecological industrial zone," "green residential compound," and so on, together with its strict environment- and ecology-oriented policies and rules. These achievements cannot be realized without an active role of the Chinese state and local governments. The lessons from Jiaxing's experiences are a clear indication of the important role of state, especially the local government, in not only promoting economic growth but also creating a sustainable economy and community. This is because the Chinese state and local governments are both the initiators and the implementers of the plans and policies of sustainable development.

Achievements of Jiaxing's Eco-civilization Construction

The basic principle of an ecological economy concerns the amount of material production that must be limited within the scale of ecosystem. In other words, human material production cannot harm the health of an ecosystem. In 2007, the Chinese President Hu Jintao clearly stated in his report to the 17th national congress of the Chinese Communist Party[1] that China should build an economy based on the

---

[1] It was officially mentioned in the speech by President Hu Jintao to the 17th National Congress held on October 15, 2007 [2], "Hold High the Great Banner of Socialism With Chinese Characteristics and Strive for New Victories in Building a Moderately Prosperous Society in All Respects."

eco-civilization, forming an industrial structure, development mode, and consumption pattern that feature energy conservation and ecological environment protection. It is the first time that the Chinese government has highlighted eco-civilization as an important strategic mission after launching the campaigns of material, mental, and political civilizations as well as a harmonious society. Eco-civilization construction will undoubtedly have significant impacts on sustainable development.

Eco-civilization construction started early in Jiaxing. Ever since China's reform and opening up to the outside world in the late 1970s, Jiaxing has followed a traditional industrial mode which mainly depended on material consumption. This kind of developing mode had indeed achieved greatly in social and economic development and accelerated Jiaxing's industrialization and urbanization. However, it inevitably brought a number of problems in ecological and social environment. As a result, Jiaxing began to realize the importance of eco-civilization construction. The government decided to build Jiaxing into a green and watery city, which claims that Jiaxing should seek not only a high level of productivity, but also a sound and peaceful living environment for the residents in the process of transforming economic growth methods and upgrading economy.

A series of measures have been taken in Jiaxing to arouse the citizens' consciousness of environmental protection, actively building an energy-saving and *environment-friendly* society, making joint efforts to be granted the title of "National Model City of Environmental Protection" and "National Ecological City and China's Habitant Prize." In the past few years, Jiaxing has endeavored to develop the recycling economy, innovate ecological environment mechanisms, take the lead in China to establish the emission right market transaction and compensation methods, try the cross-regional water treatment, construct environmental function regions, and advance the construction of provincial-level Forest City.

Ecological Concern and Environmental Protection

Jiaxing attaches great importance to ecological construction as made evident by its transformation into an ecological city as well as its limiting pollution discharge and emission reduction. Its subordinate local governments are required to take the task of "building an ecological city" as the top affair in the government annual agenda. All the subordinate counties (cities and districts) seriously carry out ecological construction by making plans and standards and coordinating regional development in terms of environmental protection and eco-civilization. By the end of 2008, five counties (subcities) have successfully been listed in "China's Ecological Demonstration Zone." Seven villages and towns have passed the preliminary examinations of "Zhejiang Ecological Villages or Towns." Five villages and towns have been awarded "China's Beautiful Environmental Villages and Towns." Now in total, there are 41 provincial-level ecological towns and 13 national-level beautiful environmental villages and towns. Besides, Jiashan County and Tongxiang City have made progress in applying for Zhejiang Ecological County (city).

Fig. 15.2 Boat on water. Wu Zhen (*left*) is an old village under Jiaxing municipality. It is an ancient water town that has recently become a popular tourist destination not only because it has been included by UNESCO in the reserve list of world cultural heritages but also because the ecological environment of the village symbolizes the harmony of man and nature

The Ecological Elements in Pollution Discharge and Emission Reduction

In accordance with "The Comprehensive Work Program on Energy Saving and Pollution Discharge and Emission Reduction" issued by the State Council, Jiaxing has made efforts to control pollution discharge and emission reduction, especially in life sewage disposal and coal consumption in fire power plants. In the first 6 months of 2008, COD has been cut down by 725.6T and SO_2 by 1, 031.1T, which drop by 4.20% and 4.00% when compared with the same period in 2007 [3]. These achievements are the results of the endeavors discussed in the following sections.

Discharge and Emission Reduction Projects

In recent years, a large number of projects concerning waste water, air, and refuse disposal have been made in Jiaxing. By November 2008, 1.3 billion yuan had been invested in town-level sewage disposal installation programs, in which 100 pumping stations and a 662.41-km pipe network were completed, and 41 towns finished sewage disposal projects. Now the handling capacity in Jiaxing has reached 685,000 tons per day, which accounts for 73.6% of the total daily life sewage water (Report from Jiaxing Water Affairs Group, December 2008 [4]). Another three projects like the second phase of Jiaxing Sewage Disposal Factory, the second Tongxiang Sewage Disposal Factory, and Pinghu Dongpian Sewage Disposal Factory are under construction. 300,000 tons can be daily handled when they are completed. Meanwhile, Jiaxing and Tongxiang's refuse-burning plants have been put into use, which can treat 1,100T per day (Report from Jiaxing and Tongxiang Environmental Protection Bureau, October 2008).

Jiaxing takes the lead in installing online auto-monitoring and supervising networks for sewage discharge enterprises in Zhejiang Province. At present, 19 surface water auto-monitoring stations and 14 air quality auto-monitoring stations have been set up, all of which can monitor the air and water in the borders between the counties (cities and districts) around the clock (Report from Jiaxing Environmental Bureau, October 2008).

Another 60 million yuan has been put into building seven pollution source monitor centers and installing 186 sets of sewage auto-monitoring facilities and 47 sets of exhaust emission auto-monitoring facilities for the serious pollution sources. Besides, the government has made a plan involving 106 reduction items and put them into practice. Now 25 enterprises are taking measures to treat their waste water with advanced technology and techniques.

Ecological Economy

While focusing on building advanced manufacturing bases, ecological and highly efficient agriculture, Jiaxing has actively adjusted industrial structure, terminated backward techniques, and established the relevant standards and regulations such as "Key Techniques and Product Guiding Catalogs" and "Evaluation System for Establishing New Industries," so that they can enhance industrial development. By the end of 2007, 4 steel-making furnaces and 13 oil-refining furnaces were closed; 126 clay brick and tile yards and 61 cement vertical kilns were demolished. More than 50% of the enterprises monitored by the provincial environmental department are now conducting clean production [5].

Likewise, some measures have been taken in adjusting agricultural structure by implementing fertile soil projects and establishing various model zones from city level down to village level. Furthermore, the appropriate technology and techniques are encouraged to produce nonpollution and green food products. So far 210 provincial-level nonpollution bases, 88 green food bases, and 7 organic food bases have been set up.

Ecological and Energy-Saving Construction

Jiaxing City Construction Bureau has a close partnership with the trade association of solar energy business in order to include energy-saving systems into Jiaxing's construction projects, especially residential households in terms of solar water heating system, HSL lamps and lanterns, air source heat pump, central ventilation system, and so on. It was expected that starting from 2009 Jiaxing would reach the target of 60% energy savings in all new construction projects.

Fig. 15.3 Skyscrapers. An example of an ecologically designed residential compound. From now on, almost all Jiaxin's newly built residential areas are designed and planned with ecological and energy-saving considerations

Comprehensive Pollution Control

Water quality is closely related to 12 elements such as dissolved oxygen, permanganate, COD, ammonia, *volatile phenol*, mercury, lead, cadmium, arsenic, hexavalent chrome, and phosphorus. The monitoring results in the past years have shown that the main factors that affect Jiaxing's water environment are the super-standard of ammonia, phosphorus, dissolved oxygen, and COD. Therefore, the control of ammonia, phosphorus, and COD is reinforced, especially the overuse of *fertilizer and pesticide* in agriculture production. Prescribed fertilizers and pesticides are encouraged in order to reduce the loss of ammonia and phosphorus. Meanwhile, the existing large-scale heat power projects are required for desulfurization.

Public Supervision

Public supervision has a unique function in the process of democracy. Therefore, the local government has issued a document titled "Proclamation for Poor Credit

Enterprises in Environmental Protection in Jiaxing." In 2008, dozens of enterprises that violated the relevant laws and regulations were publicized. Among them 8 are listed in poor credit enterprises and 43 are listed in serious environmental-polluted enterprises. All are under the public supervision.

The "Environmental Protection for Another 3 Years" Campaign

According to the "Environmental Protection for Another 3 Years" campaign issued by the provincial government, Jiaxing municipal government has initiated its own campaign – "Implementation Suggestions of Jiaxing Environmental Protection for Another 3 years" [6], setting up the key pollution problem supervision system for emphasizing the monitoring of *livestock and poultry raising* pollution:

(1) *Rural pollution control*: Taking *livestock and poultry raising* pollution and rural life sewage as key points, Jiaxing's rural environment has improved a lot. Thirty-eight raising farms are shut off and 24 raising farms are moved. 7,427 raising households are shut out of the forbidden raising areas. The reduction rate has reached 36%, and the waste disposal rate in the large raising farms accounts for 78% [7]. With the mode of "household reservation, village collection, town transportation, and county treatment," Jiaxing has basically accomplished the rural life waste collection system, covering all the villages in the city. As for sewage water treatment, Jiaxing has installed the sewage pipes in some villages. All village clinics have disposed their medical wastes through a nonpollution network.

(2) *Key city-level environmental problems*: In order to solve the most serious environmental pollution problems, Jiaxing has listed 11 key city-level environmental controlling spots with key supervision: time-limited treatment and *dynamic management*. A number of relevant documents and measures have been taken, and some achievements have been made. For instance, the air pollution caused by Jiashan waste copper wire burning and the pollution in Haiyan *fastening devices* park have significantly improved.

(3) *Industrial waste and pollution treatment*: Jiaxing has 13 provincial-level development zones. Environmental protection is very important; therefore, the government has worked out the "Development Zone (Industrial Park) Environmental Protection Layout" [8]. According to the document, by the end of 2008, at least one industrial park should be finished; in 2009, 50% of the development zones or industrial parks will be accomplished. The rest should be completed in 2010.

Besides, Jiaxing has established "Jiaxing Air Pollution Control Implementation Plan in 3 Years" in order to completely solve the problem by laying out "smoke and dust control zones," "noise forbidden areas," and "burning forbidden areas," focusing on traffic noises, entertainment noises, catering oil smoke, and small-scale coal furnace pollution.

(4) *Watercourse regulation and afforestation*: By the end of August 2008, Jiaxing had constructed 3,536 km clean water courses and cleaned 6,135 km muddy courses. The whole city had finished planting 6.71 million trees and building 3,713 ha of ecological public forest. This is why Jiaxing is known in China as one of its "green cities" [9].

New Environmental Protection Mechanisms and Public Ecological Awareness

(1) *Emission trading system*: Jiaxing has been conducting emission trading system[2] since November 2007. Nowadays, emission trading institutes are distributed throughout the city. The state and provincial departments offered 9.5 million yuan as supporting funds for emission trade. Jiaxing Commercial Bank has signed an agreement on depositing 22 million yuan with five local enterprises. One hundred and thirty-six enterprises have obtained main *pollution emission index* through emission trade. The emission trade has successfully entered the market with a total transaction volume of 100 million yuan.

(2) *Flight monitoring system*: Jiaxing municipal government has also issued "Environmental Protection Flight Monitoring Methods" in order to enhance the routine work. Since July 2007, surprise checks have been conducted twice a month in serious environmental pollution enterprises. The monitoring rate has risen from 56% in 2007 to 90% at present. This practice has effectively controlled illegal discharge.

(3) *Law enforcement against illegal discharge*: Jiaxing has issued a series of legal documents to control illegal discharge. Since 2007, the related departments such as Jiaxing *Bureau of Inspection*, Jiaxing Environmental Protection Bureau [10], Jiaxing *Public Security Bureau*, Jiaxing Intermediate People's Court, and Jiaxing *Prosecutorial Office* have jointly taken law enforcement operations. Furthermore, the economic, technological, and financial departments also join in the operation. If the enterprises are found to be illegally discharged, they will be severely punished either legally or financially. By the end of September 2008, 963 cases against environmental protection have been adjudicated with penalties of 24.57 million yuan.

(4) *Green credit experiment*: Since 2008 "green credit" has been experimented all over the city in order to set up environmental protection access. Through the green credit experiment conducted by the local environmental protection bureaus and banks, the high pollution and energy consumption enterprises have little chance to expand.

---

[2]Emission trade system was introduced in the USA, in which emission can be traded on the market. For instance, if A enterpirse treats 1 tons of SO_2 at a cost of 1,000 yuan, and B enterprise does it at a cost of 2,000 yuan, B can buy the emission from A enterprise at a price of 1,500 yuan and save 500 yuan.

Targets and *Measures* of Jiaxing's Ecological Sustainable Development

Under the general strategic targets of economic and social development, Jiaxing has actively accelerated the process of eco-civilization construction by upgrading products, adjusting industrial structure, and setting the systems and mechanisms concerning environmental protection and ecological construction so that an energy-saving and environment-friendly society will be formed. The main targets to be accomplished are as follows [11].

Main Targets

By 2012, the environmental quality in rural and urban areas of Jiaxing will obviously be improved with better environmental protection infrastructure. Through urban and rural integration in ecological construction, Jiaxing will reach the goal to become a provincial-level environmental protection model city and take the lead in Zhejiang Province. The details are as follows:

(1) *To realize the overall pollution reduction target.* By 2010, the emission of *chemical oxygen demand* will be controlled to 30,800 tons and reduced by 51.1% compared with that of 2005, and sulfur dioxide to 46,900 tons, 15% lower than in 2005.

(2) *To upgrade environmental quality.* Air quality will reach the National Second Standard, keeping more than 330 days in a year. All subordinate counties (cities) meet the provincial standards for a model city.

(3) *To promote industrial structure adjustment.* The energy-saving, water-saving, and low-pollution industries will have the priority to develop. The modern agriculture characteristic of ecology and high efficiency will be developed by further optimizing foodstuff, feeding stuff, and agricultural economy.

(4) *To construct ecological river courses.* The ecological wetlands will be reconstructed to enhance the self-purification of surface water and keep the water environment and quality improved gradually and continuously.

(5) *To perfect the environmental infrastructure.* In June 2008, all the towns in Jiaxing have built sewage treatment centers and sewage collection pipe networks. By the end of 2009, medical and hazardous wastes disposal centers will be set up and put into use. By 2010, *the sewage treatment* rate will reach 80%, and the household waste treatment rate will increase by 90%. The *comprehensive utilization ratio* of *industrial solid wastes* will rise to 93%. The harmless treatment of *industrial hazardous waste* and medical waste disposal ratio will be more than 98% [12].

Measures

In order to realize the above-mentioned goals, measures must be taken to deal with environmental pollution.

(1) *Water environment*: The government will further enhance the relevant programs and mechanisms necessary to keep the water clean and safe. First, the "Water Contamination Control Program" will be conducted. According to this program, one wetland park, one wetland reserve district, two wetland preservation projects for demonstration, and four small wetland reserve areas in Jiaxing region will be constructed so that the water environment can be improved efficiently. Second, "Water Quality Control System" and "Responsibility System" will be set up. Third, "Protection Plan for Drinking Water Sources" will be drawn up, in which the pipe network of main water supply and loop network for connecting the different regions will be constructed and the old pipe network both in the city and in rural areas will be renewed within 5 years. Fourth, "A Warning Mechanism for Drinking Water" will be established to efficiently monitor algae and toxic organic contaminants.

In addition, some measures will be taken against flood disasters. According to the plan, by 2010, all city river banks shall meet the specified standards which can withstand the big floods within 10 to 100 years. The land subsidence will also be specially cared for. The subsiding rate will decrease by 20% annually. The exploitation of underground water will be strictly controlled and totally prohibited by 2010 [13].

(2) *Industrial pollution control*: Industrial pollution is one of the most serious pollution problems. Therefore, it is of great importance to be emphasized. Several steps will be adopted in the coming years:

    (i) To enhance the renovation in highlighting pollution-causing industries and enterprises, setting renovation deadlines for unstable contaminant-releasing companies and suspending or canceling their business licenses if they fail to act. By 2010, the qualified rate shall not be lower than 90%.

    (ii) To further expand industrial structure adjustment, especially in industries like chemical, pharmaceutical, electrical, tanning, printing and dyeing, smelting, and papermaking.

    (iii) To strengthen censorship for clean production by taking enforcement measures for the enterprises which discharge the contaminants surpassing the standards. By 2010, the water-consuming amount in industry-added value of 10,000 yuan will decrease by 30% when compared with 2005, and the water-recycling rate in large-scaled enterprises will reach 70%.

    (iv) To develop the recycle economy with full efforts by focusing on the experimental units like some towns, industrial parks, and enterprises; stressing the censorship of clean production in the industries like chemical, pharmaceutical, steel, building materials, tanning, textile, papermaking, and timber; and encouraging companies to apply for ISO 14000 certification. Besides, high-tech industries like electronic information, bio-pharmaceutical, new material, intelligent instruments, and environmental protection equipment will be introduced and developed. By 2010, the output of high-tech industries will take a proportion of more than 12% of the total industry-added value.

(v) To accelerate ecological construction in industry parks by integrating every element in the park to form a resource-recycling industry chain, setting up the access in land, energy, and water resource for the enterprise to enter the park. By 2010, one provincial-level ecological industry zone, three provincial-level ecological agriculture zones, 100 recycling economy enterprises, and 20 recycling economy industrial parks will be accomplished [14].

(vi) To strengthen the prevention of nuclear radiation by focusing on ionizing radiation from nuclear facilities, wireless radio and television communication, and electric power; enhancing the facilities management of these three major electromagnetic radiation industries; and intensifying the supervision for imported waste materials and radiation sources so that the radiation environment in urban and suburb would be secured. Additionally, the emergency mechanisms of all levels will be enhanced to efficiently deal with emergent cases.

(vii) To reinforce prevention of air and noise pollution by limiting motor vehicles' exhaust gas, expanding "smoke and dust forbidden area," "noise forbidden area," and "no flame area." By 2010, the index of air pollution will be less than 80, the average annual value of sulfur dioxide and sniffing particles will be under 50 and 60 mk/m, and fine weather will have more than 330 days a year. The environmental noise will be less than 55 db (Report from Jiaxing Environmental Protection, 2008).

(3) *Agricultural pollution control*: First of all, ecological husbandry models like "Integration of Farming and Husbandry," "Synthesized Utilization," and "Biochemistry Treatment" will be constructed by the encouragement of raising herbivorous animals and special husbandry. By 2012, the proportion of raising pig, poultry, sheep, and long-haired rabbit will reach 70, 80, 50 and 50%, respectively (Report from Jiaxing Agriculture Economic Bureau, 2008). A preliminary layout of modern husbandry system will be completed. Second, standard agriculture system will be established. By 2012, 300 provincial harmless product-manufacturing bases will be built with a total area of 200,000 acres, and 390 green and organic products will pass the relevant certification [15]. Third, fertilizer will be used according to the "Less Dosage and More Efficient" project. By 2012, 10,000 samples of soil, fertilizer, and plants will be checked every year. Eighty percent of the land will be fertilized according to the examination and prescriptions. Fifteen service stations will be set up for providing information and guidance. Fourth, the synthesized utilization of straw will be developed by guiding the farmers on how to use the modern fertilizing technique to improve soil quality. By 2012, seven demonstration areas will be set

up for the application of solidified straw fuel at 50,000 tons per year and 15 straw gas stations will be completed with an annual supply of 7,000,000 cubic meters (Report from Jiaxing Agricultural Economic Bureau, 2008). The synthesized utilization rate of straw will reach 80%. Fifth, aquaculture pollution will be controlled. By 2012, pearl production will be limited within 833 acres. Aquatic products like chubs, shrimps, and crabs will be raised in ecological ways. Sixth, four afforestation projects will be carried out, that is plantation along the coast, in the farmland, in ecological parks, and in ecological villages. By 2012, the afforestation will cover more than 17% of the total area. Finally, a number of irrigation works for saving water will be conducted. By 2010, the ecological agriculture experimental villages will be further expanded with green and harmless products.

(4) *Ecological villages/towns/cities construction*: According to the program of eco-civilization construction, by 2012, 300 villages, 42 towns, and all the subordinate counties (cities and districts) will reach the goals of ecological construction of city level or provincial level. Green schools of all levels will amount to 280. Thirty hospitals will become green hospitals accounting for 70% [14].

(5) *Systems and mechanisms*: To carry out eco-civilization construction successfully, relevant systems and mechanisms have to be planned and established. The systems relate to objectives, responsibility, and assessment. All of these require that the subordinate counties' (cities and districts) governments pay special attention to the environmental protection. Air quality, water quality, general control of pollution, and some serious pollution problems are listed in the responsibility and assessment systems. If an official fails to complete the targets in due time, he/she will be criticized and forced to finish them in limited time. If he/she neglects his/her duties to cause serious environmental pollution, he/she will lose his/her post and even be prosecuted. Mechanisms include the following [16]:

(i) Emission trade system, in which COD and SO_2 can be discharged in marketing way.

(ii) Joint operation, in which the relevant departments such as the inspection bureau, environmental protection bureau, public security bureau, and intermediate court are required to take joint operations for environmental protection.

(iii) Environmental protection education, in which all the residents will be educated through schools, training courses, radio, TV, Internet, and other possible ways. Besides, the residents have the right to report to the government whenever or wherever they find serious pollution problems.

(iv) Regional coordination, in which the environmental protection departments of all levels can share information, establish the warning system, jointly inspect the water between the borders, and settle disputes.

(v) Ecological compensation, in which a principle has been determined, that is, "He who explores it has to protect it; He who destroys it must restore it; He who gets profits must pay for the losers." For instance, those who live and work in the upper reaches of a river should compensate for the lower ones.

(vi) Investment in environmental protection and eco-civilization construction needs large sums of money. How to introduce the investment is essential. One effective way is to conduct it on market by attracting all kinds of investments and regulating the prices and revenues. Another way is to enlarge the fiscal budget. Still another way is to perfect the awarding and penalty mechanisms. Last but not the least, service centers should be established so that the environment facilities will be marketed industrially, socially, and professionally.

(vii) Technology support is necessary to renovate the environmental protection technology. The government, enterprises, and the whole society should share the responsibilities and make joint efforts to create a good atmosphere. Considering the current situation in Jiaxing, the following measures can be taken:

(i) *To enhance cooperation among the enterprises, colleges, and research institutes in Jiaxing.* With the help of Zhejiang Qinghua Yangtze River Delta Research Institute, Jiaxing Applied Technology Research and Transformation Center of China's Academy of Science, and Jiaxing University, enterprises can set up their own technical research centers to develop advanced eco-environmental protection technologies and speed up the transformation of hi-tech achievements. The advanced eco-environmental protection technologies and management modes from abroad should be introduced so that they may upgrade the advanced facilities and technologies in a number of enterprises. On the other hand, the technology in energy saving and high efficiency should be further developed to form an industrial chain so that Jiaxing will have the technological advantage in industrial waste treatment, agricultural pollution, and eco-restoration.

(ii) *To further accelerate the construction of environmental protection auto-inspection networks and institutions.* On the one hand, the auto-inspection networks should be connected with the provincial network to deepen the information platform. On the other, all the pollution enterprises should be listed in the inspection network for self-inspection according to the requirements set by the environmental protection departments of all levels.

(iii) *To build the technological inquiry service platform.* The relevant guilds and the energy-saving technological service agencies should be fully developed so that they can play an important role in the construction of eco-environments and urbanization by providing the advanced scientific and technological services.

Lessons from Jiaxing Experience: Chinese Social Capitalism

Jiaxing's achievements in ecologically and environmentally sustainable development provide an empirical account of the success of the Chinese development model of social capitalism [17]. The Chinese "socialist market economy," a sinicized term of social capitalism, is a distinct form of capitalism characterized by active state intervention and close state–market relations. The emergence of new institutional entrepreneurs and their role in institutional innovations play a positive role in encouraging marketization and decentralization of state capacities and public resources without falling into economic and social disembeddedment. Through the case study of Jiaxing Municipality, this chapter focuses on identifying the key ingredients of the "embedded" relations between the Chinese developmental state and the market that have led to sustained economic growth and will eventually lead to sustainable development.

Understanding the Chinese Developmental State

Largely inspired by intense curiosity on this first case of rapid industrialization outside the Western cultural sphere, Japan and the East Asian newly industrialized economies became the object of various academic studies and interpretations. The World Bank had published a special study report on *The East Asian Miracle* [18], which generated a global debate on the various factors behind the East Asian success.

The Chinese experiences can be explained by reference to the core features of the dominant paradigm for the development of what Chalmers Johnson calls the "Capitalist Developmental State" [19]. China's policy model clearly resembles the East Asian experiences: building on a strong authoritarian national leadership and an elite state bureaucracy pursuing developmentally oriented policies, including the direct role of the state in governing the market [20]. China's success in the last three decades of economic reform has been led by a strong, pro-development state that is capable of shaping a national consensus on modernization and maintaining overall political and macroeconomic stability in order to pursue wide-ranging domestic reforms. There are a few unique features of this type of developmental state which have fostered the dynamic aspects of economic growth:

(1) It sees social and economic development as the over-arching objective of government. It creates social stability and political predictability and maintains a manageable equality in distribution in order to prevent crisis between capitalist accumulation and class/sectoral exploitation. It plays a leading role in fostering, guiding, and ensuring economic growth and technological modernization over the long term.

(2) It puts forward national development goals and standards that are internationally oriented and are based on non-ideological external referents. It is eager to absorb worldwide development experiences without abandoning its own

policy-making sovereignty as to when, where, and how to adopt foreign ideas. The state is determined to play an active role in financial control over the economy even in face of international pressure to liberalize its financial sectors.

(3) It sets up an infrastructure of productive forces and labor markets which target the global market so that its export-oriented economic growth is sustainable on a long-term basis. Its national education system is also designed to serve economic growth and the foreign market.

(4) It initiates state-driven industrial policies. It recognizes and empowers a bureaucratic elite capable of administering the system without subjecting to political influence by various interest groups so that they can function professionally and independently. On the other hand, economic policy-making processes involve close government–business collaborations in order to correctly respond to market signals.

(5) It believes that free market mechanism sometimes needs explicit "administrative guidance" and "directed credit," which pick up the winners or prioritize some industries over others. Public and private sectors often work together to pursue social and economic goals. The government not only regulates business enterprises but also assists them with overheads and other preferential policies. It channels foreign direct investment to target at strategic private businesses, while business enterprises assist the government in reaching social and economic goals.

(6) It does not allow liberal ideologies to confuse the national consensus and does not permit the development of political pluralism that might challenge its goals. An inactive civil society does not intend to interrupt the consensus. It does not see Western democracy as a political system on its own that will necessarily lead to economic and social development. It believes what a country needs at its initial developmental stage is discipline more than democracy.

State Corporatism

State corporatism in China can be identified as a type of "institutional clientism" [21]. It demonstrates distinctive features of the interactive relations between private business and the state including local political structures at various levels:

- Institutional clientism implies the transformation of institutionalized social relations from monopoly to marketization of the country's resources, either through official's position or through client ties between private actors and office holders. It is an integrating process in which the control power of state institutions is incorporated into economic calculations and business activities that reflect commodification values.
- State policies and decision-making calculations on resource allocation are integrated into market competition logic in which local governments compete with

each other over mobile capital and labor resources by providing a competitive local infrastructure and business environment. Nevertheless they are centered on social relations including private businesses that seek to have a share of these resources. The state's previous direct monopoly is replaced with a new regulatory monopoly aiming at facilitating and constraining market interactions such as licensing, quota allocations, and so on. This type of state–market clientelist relations can promote efficiency in an emerging market because it permits long-term calculation rather than short-term speculation.

- Institutional clientism involves a reconfiguration process in which state politics are more directed toward market competitive ends. However, the new market system cannot function independently from the political system in which the legacies of the Communist Party and state create and constrain processes of cooperation and competition. This is because institutions are not neutral, and they are culturally and socially conditioned. The Communist Party has been developing strategies in order to cope with the new changes. For example, the party is willing to co-opt new party members including the new capitalists and create new links with other emerging organizations.

- Institutional clientism is maintained through an institutional framework in line with *social trust*. The new clientelist relations promote state–market cooperation, thus avoiding an "either–or" situation, that is, either the politics politicizes the market or the market marketizes the politics. In other words, the market itself has no objection to political authoritarianism as long as it cooperates with market mechanism. The emphasis of clientism is aimed at the maintenance of social order and political stability, and such expectation has to be understood and incorporated into market interactions.

- Since the patron–client relationships are based on social trust that reflects the interests of both sides, private business is keen on ties with officialdom. In this type of relationship, "power is embodied not only in the monetary gains derived from trade but also in position in network. Diffuse forms of social, symbolic, and cultural capital shape relative resources and outcomes in interpersonal bargaining" [17:31]. To do business is understood not solely as utility maximizing in market transactions but as cultivating the personal and social relations (in Chinese "*cultivating Guanxi*"). The Chinese party-state likes to dine and dance with private business so that it is able to continue to control the politics while becoming enriched by the market.

- The public–private clientelism unveils that formal legal property rights and definitions of an individual entrepreneur are less important than the "social environment" in determining the outcomes of business activity and market performance. In other words, having a good relationship with local party and state officials is much more central for doing business than the formal ownership classification of that enterprise. Commercial rationality in China is less determined by relying on market opportunities than by cultivating strong ties with "Guanxi," which in turn will facilitate those market opportunities.

Lessons from the Chinese Model

Some concrete lessons can be learned from this model. The first is *selective learning* in which China has been trying to selectively borrow from the neoliberal American model that emphasizes on the role of the free market, innovation, entrepreneurship, and international trade. The second is *trial adaptation*, in which the state has been following the carefully planned sequencing and priorities: less controversial reforms first, more difficult ones second; rural reforms first, urban ones second; policy preference to coastal areas first, inland second; economic reforms first, political adaptation second. The advantage of Chinese practices demonstrates that the experiences gained in the former create conditions for the latter. Seen from these perspectives, Jiaxing is a good example – already the surrounding cities of Jiaxing have started to localize Jiaxing experiences of development into their own development projects and planning.

Conclusion

It is true that Jiaxing has made great achievements in environmental protection and ecological environment in the past few years. However, there is still a long way to go. Eco-civilization construction is a comprehensive project which needs persistent care and participation from all governments, enterprises, and residents so that the goal of sustainable development will be reached and that an energy-saving, environment-friendly, and harmonious society will be finally formed.

Because Jiaxing is only a medium-sized city in China, many aspects of its success can be assessed only at the local level. However, the experience of Jiaxing provides the empirical framework of understanding Chinese social capitalism [17], manifested by the active role the Chinese developmental state and local governments as well as the unique state–market relations. To put it more clearly, the post-Mao transitional market economy along with the fundamental institutional transformations characterizes a distinctive style of capitalism in which the marketization of the former command economy, the active role of the Chinese party-state and local government, the variety of forms of property and business ownership, the traditional culture of clientele-based social relations, the institutional legacies of socialism, and the emergence of market-based institutions all provide rich empirical context to conceptualize and theorize the "embeddedness" characteristics and the new socio-institutional hegemony in post-reform China.

It also needs to be pointed out that Jiaxing is one of many developed cities together with other rich cities along China's southeast coastline. The achievements Jiaxing has accomplished, the measures Jiaxing has taken, as well as the targets Jiaxing intends to reach cannot be generalized, nor can they be compared with other municipalities in less developed regions. Due to China's internal gaps in living standards, economic and social developments between different provinces and regions and priorities and levels of ecological and sustainable developments in various parts of the country are also very different. The difficulties China is facing concerns how to balance the tradeoff between rapid economic growth and rising living standards on the one hand and ecological, environmental, and energy sustainability on the other.

References

1. Jiaxing Statistics Monthly Report, Oct. 2008
2. President Hu Jintao's Report in the 17th National Congress of CPC, Oct. 2007
3. Jiaxing Environmental Protection Monthly Report, Oct. 2008
4. Jiaxing Water Affairs Group Report, Vol. 11, Dec. 2008
5. Annual Report of Jiaxing Development & Reform Commission, 2007
6. "Implementation Suggestions of Jiaxing Environmental Protection for Another 3 Years", Zhejiang Provincial Government Documents Vol. 3, Mar. 2007
7. Annual Report of Jiaxing Agricultural Economic Bureau, 2007
8. "Development Zones (Industrial Park) Environmental Protection Layout", Jiaxing Municipal Governments Vol. 10, Oct. 2007
9. Monthly Report of Jiaxing Water Conservancy Bureau, Sept. 2008
10. Report of Jiaxing Environmental Protection Bureau, Vol. 11, Nov. 2008
11. Annual Report of Jiaxing Development & Reform Commission, 2008
12. "Jiaxing Sewage Treatment Projects", Jiaxing Environmental Protection Bureau, 2008
13. Jiaxing Water Conservancy Bureau Report, Vol. 10, Oct. 2008
14. Jiaxing Development & Reform Commission Report, Vol. 10, Oct. 2008
15. Annual Report of Jiaxing Agricultural Economic Bureau, 2008
16. Jiaxing Environmental Protection Report, July, 2008
17. Woodrow W. Clark and Li Xing, "Social Capitalism: an economic paradigm for the transfer and commercialization of technology", *International Journal of Technology Transfer and Commercialisation*, Vol. 3, No. 1, 2004
18. World Bank, *The East Asian Miracle: Economic Growth and Public Policy*. New York: Oxford University Press, 1993
19. Chalmers Johnson refers it to Japan together with South Korea, Singapore, Taiwan and Hong Kong, see his books (1982) MITI and the Japanese miracle: *The growth of industrial policy, 1925–1975*. Stanford, CA: Stanford University Press; and (1995) Japan: who governs? The rise of the developmental state. London: Norton
20. Robert Wade, *Governing the Market: Economic Theory and the Role of Government in East Asian Industrialization*. Princeton: Princeton University Press, 1990
21. David L. Wank, *Commodifying Communism – Business, Trust and Politics in a Chinese City*. Cambridge: Cambridge University Press, 1999

Chapter 16
Japanese Experience with Efforts at the Community Level Toward a Sustainable Economy: Accelerating Collaboration Between Local and Central Governments

Kentaro Funaki and Lucas Adams

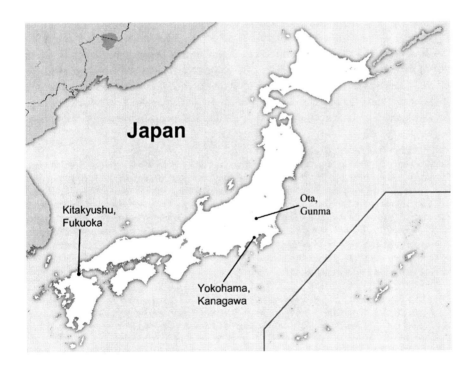

K. Funaki (✉)
Japan External Trade Organization, Los Angeles, CA 90017, USA
email: kentaro_funaki@jetro.go.jp

W.W. Clark II (ed.), *Sustainable Communities*, DOI 10.1007/978-1-4419-0219-1_16,   243
© Springer Science+Business Media, LLC 2010

Introduction

Living in a country with limited natural resources and high population density, the people of Japan had to work on sustainability throughout their history as a matter of necessity. With scarcity of arable land – some 70–80% of the land is mountainous or forested and thus unsuitable for agricultural or residential use – people clustered in the habitable areas, and farmers had to make each acre as productive as possible. The concept of "no waste" was developed early on, as a particularly telling, literal example; the lack of large livestock meant each bit of human waste in a village had to be recycled for use as fertilizer. Along with creating this general need for conservation, living in close proximity to others inspired a culture in which individuals take special care in the effect their actions have on both the surrounding people and the environment. As such, a desire for harmony with others went hand in hand with a traditional desire for harmony with nature. Nature came to be thought of as sacred, and to come into contact with nature was to experience the divine. Centuries-old customs of cherry blossom or moon-viewing attest to the special place nature has traditionally held in the Japanese hearts.

However, nature took a backseat to industrial development during the drive toward modernization and economic development that began in Japan in the latter half of the 19th century. After nearly 300 years of self-imposed isolation from the world, Japan was determined to catch up with the industrialized West in a fraction of the time it took Europe and the USA to make their transitions, eventually emerging as a great power in the beginning of the 20th century. Economic development

Lucas Adams has spent several years in Japan, including time studying abroad at Waseda University, and received his BA in Japanese language and culture from UCLA. After graduation, he was selected to participate in the JET Programme, a Japanese governmental cultural exchange and teaching program, and spent 2 years working in Shizuoka before moving to Tokyo to work in business. After returning to the USA, he joined the Los Angeles office of the Japan External Trade Organization (JETRO) and worked in the technology division as project coordinator, promoting business exchange between Japan and the USA, specifically in the field of Japanese green technology. He is currently pursuing his J.D. at Georgetown University in Washington, D.C.

Kentaro Funaki is from Kanagawa, Japan, and holds an MA in international affairs from Columbia University in New York, as well as a BS in nuclear engineering from the University of Tokyo. Before coming to southern California in the summer of 2007, he spent 16 years working in the Japanese government, mainly in the Ministry of Economy, Trade and Industry (METI). His focus there was on energy policy, including renewable and nuclear power, and international trade and economic development. He was involved in drafting Japan's renewable energy targets and associated policy program, including the newly adopted Renewable Portfolio Standard and a photovoltaic policy package. He was also involved in international affairs in competition policy at the Japan Fair Trade Commission for 2 years starting in 2001, where he joined the Japanese delegations to the WTO and FTA negotiations. Currently, he works in the Los Angeles office of JETRO as an executive director in the research and technology division, where he supports business partnering for clean-tech and bio-tech companies on both sides of the Pacific by facilitating initial contacts and providing advisory services and networking opportunities.

Photos courtesy of Kitakyushu City (Kitakyushu photos), Ota City (Ota City photos), The City of Yokohama (Yokohama photos)

continued unabated until World War II, when much of the capacity was destroyed by Allied bombing, but growth restarted again in the postwar period at a rapid pace. By the 1980s, on the strength of its industry and manufacturing capabilities, Japan had attained its present status as an economic powerhouse.

While this incredibly successful period of development left many parts of the country wealthy, it also resulted in serious environmental problems. In addition, the oil crisis hit Japan particularly hard because of its lack of natural resources, making it difficult for the industrial and manufacturing sectors to keep working at full capacity. To respond to the effects of pollution, municipalities began working in earnest on ways to reduce emissions and clean up the environment, while Japanese industry responded to the oil crisis by pushing for an increase in energy efficiency. At the same time, Japan's economy was evolving more toward processing and high technology, which held the promise of further increases in energy efficiency. In this way, the Japanese government, industry, and academia have worked hand in hand with communities to reincorporate traditional Japanese ideas about conservation and respect for the environment in order to create sustainable lifestyles compatible with modern living.

Community-level efforts in Japan, supported by government initiatives, have led to unique advancements in energy efficiency and sustainable lifestyles, including novel ways of preventing and eliminating pollution. As it stands, Japan is responsible for some 4% of global CO_2 emissions[1] from fuel combustion, and though this is the lowest percentage among major industrialized nations, it is still something the country intends to reduce, with a long-term goal of reducing emissions by 60–80% by 2050. With the majority of energy still coming from coal,[2] Japan is also attempting a large shift toward renewable energy. This chapter will explore such

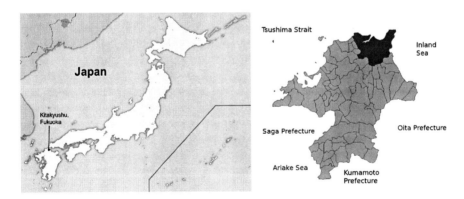

Fig. 15.3

---

[1] According to 2005 IEA figures.

[2] According to 2006 IEA figures, oil accounts for 45.6% of the total primary energy supply, followed by coal/peat at 21.3%, nuclear at 15.0%, and gas at 14.7%, with renewable sources like hydroelectric, solar, and wind still at less than 4%.

efforts toward achieving this goal in three quite different cities – Kitakyushu, Ota, and Yokohama, all ranging in size from a relatively small city to a large metropolis – and shed light on lessons the world can learn from Japan's citizens, businesses, and governments on how to work together to move toward establishing a sustainable society and addressing global warming.

Kitakyushu City: An Environmentally Friendly Industrial Center

Kitakyushu City, Fukuoka Prefecture: City Overview

Kitakyushu (literally "Northern Kyushu") is a city located in Fukuoka Prefecture in the northern part of Kyushu, the southern island of Japan. Kitakyushu City itself was established in February 1963 as an amalgamation of five previously independent cities and has a population of around a million people. Kitakyushu is part of one of Japan's four largest industrial zones, a legacy that can be traced back to 1901, when the first government-managed steelworks in Japan were established there. With its key steel, chemical, electric, and ceramic industries, the region played an important role in the modernization of Japan in the early 20th century. In fact, because of its importance as an industrial zone, the city of Kokura, which is now part of Kitakyushu, was the original target for the "Fat Man" atomic bomb, but due to poor visibility over the city, the bomb was instead dropped on Nagasaki.

However, heavy industry in the region began to decline as the source of imported material for heavy industries changed from China to Australia and the export destination changed to the USA in the period from 1955 to 1965. This decline was compounded by the energy revolution in the shift from coal to oil, which made production fueled by materials from neighbors even more difficult to maintain. At the same time, the city was confronting serious pollution still remaining from the rapid economic growth in heavy industries. Thanks to the combined efforts of citizens, businesses, research institutions, and the government, the city was able to overcome this problem and, in doing so, accumulated various technologies and experience, which it then used to reinvent itself as a center of environmental and recycling industries. As a result, Kitakyushu was recently recognized as the "Eco-City" of Japan, after winning the sixth annual Japan's Top Eco-City Contest held in 2006.[3] An important part of this environmental movement in Kitakyushu was its involvement in the Eco-Town project.

---

[3] The purpose and goal of the Japan's Top Eco-City Contest is the birth of an "Eco-City" in Japan. A network of 11 national NGOs conducts a survey of municipalities, the results are tallied, and the municipality receiving the most points is recognized publicly as the "Environmental Capital." The contest allows for a dialogue between municipalities and NGOs on approaches appropriate for their regions as well as ways to address issues stretching across multiple regions.
More information (in Japanese) at: http://www.kankyoshimin.org/jp/mission/ecocity/ecocap/index.html

The Eco-Town Project

In the 2004–2005 fiscal year, the Global Environment Centre Foundation (GEC), a nonprofit Japanese organization, implemented a research project on "Eco-Towns in Japan" in cooperation with the United Nations Environment Programme (UNEP) Division of Technology, Industry, and Economics (DTIE) International Environmental Technology Center (IETC) [1]. Eco-towns are areas where urban planning and environmental management tools are applied to pursue synergies in resource utilization, waste management, environmental preservation, and promotion of industrial and economic development. The Japanese Ministry of International Trade and Industry (now known as the Ministry of Economy, Trade and Industry, or METI) established the Eco-Town project in the fiscal year 1997 in order to promote the Zero-Emission Concept, which has three aims: first, making gross input equal gross output (in other words, eliminating waste); second, reducing greenhouse gas emissions and environmental burdens and promoting energy-saving measures; and third, fostering collaboration among collective industries in various fields as well as among administrative districts beyond their borders.

The Eco-Town project is a plan for creating new environmental towns in the 21st century and is conducted in coordination with the Ministry of Environment. Twenty-six eco-town plans have been approved and many eco-towns were developed in the last 10 years by utilizing regional technology and industry in Japan. The Eco-Town project is aimed at making every bit of waste a raw material for another type of production, with the goal being the eventual achievement of zero waste (or zero emissions) and the construction of a society that revolves around the recycling of resources.

Kitakyushu Eco-Town

In Kitakyushu, the "Kitakyushu Eco-Town Plan" – authorized by the Ministry of Economy, Trade and Industry (METI) and the Ministry of the Environment – was formulated as a basis for developing environmental and recycling industries, and the city has concrete projects ongoing in all areas [2]. The project was formulated as the "Executive Proposal for the Kitakyushu Eco-Town Plan" back in April 1998, with its development going back to October 1989. The plan set down the direction of the basic efforts, and the city is engaged in a unique regional policy that integrates environmental and industrial developmental policies.

A key fixture of the Kitakyushu Eco-Town is the Comprehensive Environmental Industrial Complex [3] (in Japanese eko-kombinaato, from the Russian word kombinat meaning "combination"), Kitakyushu's industrial region, which practices the cyclical use of energy and materials – a cycle between raw materials, manufactured goods, and waste – along with having a system in place for cascade use. In the Comprehensive Environmental Industrial Complex, energy, resource, and water use are not closed within a single industry, but are used with versatility between factories, between different industries, and from industry to consumer. Due to the continued pursuit of efficiency in the use of energy and resources, Kitakyushu City has been able to increase economic competitiveness in industry and improve civic life through the reduction of waste and CO_2, establishing a city in which industry and the community in the entire region are united in an effort to coexist peacefully with nature.

Fig. 16.1 Comprehensive Environmental Industrial Complex, Hibikib Recycling Area, Kitakyushu City, Fukuoka, Japan

Other Ongoing Efforts in Kitakyushu City

Kitakyushu also has a variety of its own independent ongoing efforts to create a sustainable community. One such effort is Eco Premium (*eko puremiamu*) [4], a project in which goods, technologies, and industrial activities that feature a low environmental load are selected from Kitakyushu and receive a special designation of "Eco Premium." The selected goods and services are then used to support the public relations efforts of Kitakyushu City. Benefits include certificates and stickers recognizing the award, exhibition in environmental trade fairs, and inclusion in Kitakyushu Eco Premium catalogs.

Another effort spearheaded by Kitakyushu is known as Eco Action 21 [5], an environmental management support system primarily geared toward small- to midsized businesses that require initiatives to reduce CO_2 and waste and the creation

and announcement of environmental practices reports. Eco Action 21 provides certification and registration to verify environmental management at a fraction of the
cost of ISO14001, the standard international specification for an environmental
management system, which encourages the participation of businesses of all sizes.

Kitakyushu is also one of several sites for a Next-Generation Energy Park [6],
a project METI initiated to expose the public to next-generation energy technology like sun power and increase understanding through direct experience. Local
governing bodies will implement plans as they feel appropriate. Nationwide, METI
recognized six plans in 2007 and seven in 2008. One example in Gobo City has
exhibits on solar, hydroelectric, biomass, and wind power, along with solar-powered
cars that visitors can drive around a raceway. Another in Maniwashi City is called
"Biomass Town" offering a tour that runs through exhibits on all the different stages
that go into generating power from biomass. Other parks are planned for or completed in Genkai Village, Nagasaki Prefecture; Sapporo City; Rokkasho Village;
Ota City, Gunma Prefecture; Yamanashi City, Aichi Prefecture; Izumo City; and
Anan City. The Kitakyushu Next-Generation Energy Park showcases various installations related to energy generation in the eco-town, featuring those in the Eco-Town
Center – like the solar power generation and fuel cell exhibition hall – as focal
points.

The experience of Kitakyushu demonstrates how, if citizens and government
can make a comprehensive plan with companies, a city that hurt the environment
through its old industries can work to transform itself into one that helps save the
environment with its new industries.

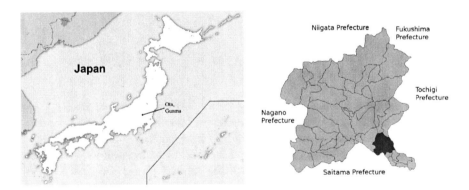

Ota City: Solar Town

Ota City, Gunma Prefecture: City Overview

Ota is a small city located some 50 miles northwest of Tokyo in eastern Gunma
prefecture. It has a population of around 230,000 and, as the home of carmaker
Subaru, is the manufacturing center of the prefecture. Like Kitakyushu, Ota City in
Gunma Prefecture sports a Next-Generation Energy Park, and it has also worked

proactively to promote the installation of photovoltaic systems in the city through subsidies, even remodeling the city hall in 1998 to include 30 kW of solar power generation capability (including panels on the roof and amorphous solar cells in the windows). Ota City, however, is most well known for its extensive solar installations that have earned it the nickname "Solar Town."

Residential Solar Power in Japan

As of November 2008, residential-use solar power generation systems have been put in place in around 380,000 homes in Japan. A close examination of the data on shipments domestically in Japan shows that 80–90% are intended for residential use, and such shipments are likely to increase, as the government aims to have solar panel equipment installed in more than 70% of newly built houses by 2020 to meet its long-term goals for reductions in emissions. Current goals for solar power generation in Japan are to increase its use 10-fold by 2020 and 40-fold by 2030, and large proposed subsidies for the installation of solar panels – 9 billion yen or $99.6 million total in the first quarter of 2009 – along with tax breaks for consumers will continue the acceleration of solar adoption by Japanese households [7, 8]. In recent years, in places like Europe, China, and Taiwan, tremendous growth has been seen in energy generation entirely from solar power installations, but those have mostly involved large-scale power facilities not fit for individual households. In Japan, however, as solar power generation systems for residential use become increasingly commonplace, they are likely to become concentrated in communities, and the possibility exists of heretofore unseen problems.

Pal Town Josai-no-Mori

With this danger in mind, the New Energy and Industrial Technology Development Organization (NEDO) – a Japanese governmental organization – sought to study this potential trouble and entrusted the electric company Kandenko Co., Ltd., with the central role in its "Demonstrative Project on Grid-Interconnection of Clustered Photovoltaic Power Generation Systems." The research focused on what happens when solar power generation systems for residential use are concentrated in one area, investigating the effects on the distribution system and technologies to use in response [9, 10].

Ota City in Gunma Prefecture was chosen to be the site of the research project [11], based on its geographic features – flat land and long hours of daylight – and preexisting local support for solar technology. The project was called "Pal Town Josai-no-Mori". As a result of the research project, 553 houses in total have had solar panels installed on their roofs, generating a total of 2129 kW, which qualifies as a "mega-solar" facility. The experimental study in Pal Town ended in March 2008 and has allowed for concrete and detailed insight into what sorts of problems arise in such an installation and the correct responses to the said problems. The research centered on two main issues: the rise in system voltage in concentrated residential solar facilities and the development of "new islanding detection equipment" that would reliably and automatically switch off the photovoltaic system when power outages occur.

When electric power is sent from a transformer substation to a household, the voltage drops over the distance it travels through the power lines. Power companies make allowances for this and use voltage management systems to control the amplitude of the voltage on the power lines, but the assumption in this system is that power is only going to travel one way – from the power company to the household – and thus no allowance is made for the reverse power flow of surplus energy generated from a residential solar installation. As residential solar becomes increasingly popular, even if each system has a generation capacity of only 3–5 kW on average, when hundreds of these systems are concentrated in a single area, the power generated in the town would reach "mega-solar" capacities of and above 1 MW, which would likely strain the existing power distribution system. On a nice, sunny day, the surplus power generated could easily exceed the upper limit for the voltage management system.

Fig. 16.2 "Pal Town *Josai-no-Mori*", Ota City, Gunma, Japan

Fig. 16.3 "Pal Town *Josai-no-Mori*", Ota City, Gunma, Japan

Fig. 16.4 "Pal Town *Josai-no-Mori*", Ota City, Gunma, Japan

For this reason, ordinary solar power systems are equipped with voltage control systems that are designed to automatically keep the voltage from exceeding this upper limit. This would seem to solve the problem, except that it presupposes a small number of installed solar power systems. In concentrated installations, the voltage control systems of each solar power generator often operate too frequently, causing some generators to shut off and decreasing the amount of power generated. In the end, one can end up with a solar power system that does not generate any power on a sunny day – negating its entire purpose.

To solve this problem, technology to eliminate restrictions on photovoltaic system output to avoid the shutoffs in operation was developed in the course of the research project. Installed in all the houses in the town, the equipment uses batteries to store any surplus energy generated, thus controlling the rise in voltage. In each house 16 batteries were installed, with a total of 9.4 kW of storage possible. Using this innovation, electricity can continue to be generated and stored even on the sunniest of days.

The second issue addressed through the research project was the need for systems that would effectively cope with power outages. In storms or other conditions that lead to cut power lines, a power company will stop power transmissions from transformer stations completely to protect people from receiving electric shocks from the cut lines. Similarly, residential solar power systems are equipped with systems for shutting off one's own power in the event of a power outage. The only problem is the lack of precedent showing that this will work reliably when these residential systems are combined to form a concentrated installation. If for some reason power could not be shut off, accidents could potentially occur.

In the course of the research project, equipment for reliably shutting off the system that also operated without malfunction in ordinary situations was developed. Using in-depth data, analysis, and evaluation, combined with the lack of a single major issue during the research period, researchers in Ota City were able to prove the effectiveness of their equipment. One major reason solar power systems have not been installed in concentrated regions up to now was the lack of acceptance by power companies, which were worried about liability for accidents. Even developers who might have wanted to install concentrated solar installations had to give up on the idea for this reason. The results of the research project in Ota City should eliminate this problem and leave the door wide open for the rise of concentrated solar installations in the future.

The original research project has ended, but the next step will be another NEDO project under more general conditions to promote the spread of the results of the concentrated photovoltaic system research. Interestingly, the research project has inspired a new movement in Ota City itself toward solar power, with new buyers of residential property wanting to install solar panels in homes – even at their own expense. Though Ota City was branded as the "Solar Town" based on its participation in a government research project, it seems likely to earn the moniker through continued acceptance of solar technology by the local population.

Other Ongoing Efforts in Ota City

Aided by grants from the Ministry of the Environment, Ota City was also a co-partner in the "model city project for a virtuous circle between the environment and economy," which ran for 3 years beginning in 2004. This project was about taking the initiative to improve public facilities and schools to make them more energy efficient. It aimed at preventing global warming by supporting the efforts of homes and businesses to become more energy efficient and have a positive effect on the economy through investments in new green technology. Through the project, 20 public facilities were renovated to make them more energy efficient, leading to a reduction of around 1000 tons of CO_2 emissions in 1 year, along with a savings of 35,000,000 yen (approximately $385,000).

The "Super Eco House" [12] was also built in Pal Town, increasing public awareness of the new energy-efficient technology available for the reduction of CO_2 emissions.

As previously noted, Ota City is also home to an Energy Park, with its plan for the Ota City Next-Generation Energy Park approved in June 2008 by the Agency for Natural Resources and Energy of METI. The park was designed to be a fun place for people of all ages to see, touch, and experience new energy technologies, with the added intention of raising awareness and spreading knowledge among the local community as well as the rest of the country about the energy-related installations within the city.

Thanks to the cooperation of the citizens and local officials of Ota City, a government organization has been able to demonstrate the viability of technologies with enormous promise for creating sustainable communities, in the process inspiring citizens to independently choose new, greener lifestyles.

 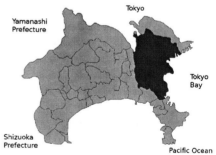

Yokohama: A Green Metropolis

Yokohama, Kanagawa Prefecture: City Overview

Yokohama is the capital of Kanagawa Prefecture and a major commercial hub of the Greater Tokyo Area. One of Japan's major port cities, it has a population of 3.65

million – making it Japan's largest incorporated city[4] – and a total area of 437 km². Yokohama participates in environmental programs on a national and local level through its selection as an Environmental Model City and through the development of several of its own programs for achieving sustainability.

The Environmental Model Cities Program

The Environmental Model Cities program [11] was put together by the administration of Prime Minister Fukuda, who gathered a team of experts to cull opinions on how to plan cities that would effectively combat global warming through the reduction of carbon emissions. Cities across Japan submitted proposals to the prime minister hoping to be designated as an Environmental Model City. Of the 81 proposals received, many had ambitious goals of reducing greenhouse gas emissions, but some were truly groundbreaking, including plans for making the region carbon-free, counter plans that would address global along with local emissions, suggestions for economical changes in the actions of citizens and structure of the cities in which they live, and plans that offered ways to stimulate the economy. The 81 cities that submitted proposals were evaluated on their fulfillment of several demanding criteria.

First, the proposals needed to include drastic cutback targets, with greenhouse gas emissions cut in half by 2050 and energy efficiency improved by more than 30% by 2020, with the expectation of exceeding these goals. Second, the proposal needed to demonstrate leadership on the part of the city by including innovative approaches for reducing greenhouse gas emissions, especially integrated efforts and those that would improve the general quality of life for residents. Third, the proposal had to stress regional adaptability, advocating efforts that adapt to and take advantage of the particular characteristics of a region. Fourth, the proposal had to pass a test of feasibility, with the emphasis toward plans that included the participation of citizens, industries, universities, NGOs, and other groups and that had realistic, attainable goals. Lastly, the proposals had to be durable, meaning they would have frameworks in place that encourage continuing participation and instruments to urge the spread of the plan onward, including education for the next generation about environmental efforts and long-term efforts to ensure development.

Six cities were judged to have fulfilled the criteria and were selected as Environmental Model Cities. Two were the large cities of Yokohama and Kitakyushu, two the provincial cities of Obihiro and Toyama, and two the small towns of Shimokawa and Minamata. In selecting these proposals built on agreement at the local level between the governments, citizens, and industry of the region, the prime minister hoped they could be the beginnings of a mighty swell in sustainability that could sweep the entire country.

---

[4] Tokyo is considered a "Metropolis" and one of Japan's 47 prefectures, and thus is technically not a city.

Environmental Model City: Yokohama

Even before its selection as an Environmental Model City, Yokohama had already taken significant steps toward becoming more environmental-friendly through its G30 Plan [14], implemented by the city in 2003 to achieve a 30% reduction in garbage by 2010. The "G" can be taken to mean "*gomi*" ("waste" in Japanese), "garbage," or "*genryou*" ("reduce" in Japanese), and the "30" represented the 30% reduction goal. "G30" can also mean "zero waste," as "3" can be read as "*mi*" in Japanese, making the "G30" into "*gomi zero.*" The plan relied on cooperation between citizens, businesses, and government: citizens would adopt a more environmental-friendly lifestyle and actively sort their garbage, businesses would design and produce products which reduce waste and recycle used products, and the government would create systems for the three Rs along with raising awareness and monitoring and exchanging information on the program.

Fig. 16.5 Shinmeidai Landfill, City of Yokohama, Japan

Household waste was reduced through an increase in the items sorted for collection. In the past, there were 5 types and 7 items, but the G30 plan created 10 types and 15 items. Concerted governmental efforts to inform citizens about the plan included briefing sessions at neighborhood associations, campaign events at train stations and shopping centers, and early-morning guidance sessions to answer questions about garbage sorting. The inspection of collected garbage at incineration plants became stricter starting in December 2003, with garbage containing large quantities of recyclable waste sent back to the source. G30 seminars were held at elementary and junior high schools, and students and other residents were given tours of the incineration plants.

The program was a substantial success from both an environmental and a financial standpoint. On the environmental end, the CO_2 emissions resulting from garbage treatment were significantly reduced: the reduction in the 2007 fiscal year was 840,000 tons compared to 2001 levels – a reduction equivalent to the CO_2 absorbed by 60 million Japanese cedars in 1 year. Additionally, the 30% garbage reduction target was achieved in 2005 – 5 years earlier than planned. On the financial end, the city realized a significant benefit from the plan, as the substantial reduction in garbage made it possible to close down two incineration plants, saving a future cost of 110 billion yen for reconstruction and 3 billion yen per year in operational costs.

As part of their proposal to be selected as an Environmental Model City, Yokohama trumpeted this "citizen power" recycling program that had achieved such dramatic and accelerated reductions in waste and emissions and also set goals for a more than 30% reduction in greenhouse gases per person by 2025 and a more than 60% reduction per person by 2050 (compared to 2004 levels). With this proposal as its guide, the city continues to strive toward achieving a big city, zero-carbon lifestyle through knowledge sharing, expansion of choices, and citizen power efforts.

One part of the proposal adopted by Yokohama promotes the construction of zero-carbon and long-lasting houses (so-called "200-year houses"), by offering financial incentives, such as lowered property taxes, to houses that exceed current environmental performance standards for buildings (like *CASBEE* [15][5]). Another program known as "Eco service" (*eko motenashi*) requires carbon offsets for concerts and sports events held by the city or at city buildings, with the goal of eliminating disposable cups and dishes at fast food restaurants and hotels.

The city also makes efforts toward energy saving in large-scale development. One such effort is the suggestion system for city planning, through which groups like owners and NPOs offer guidance on land use plans as part of the planning law for fixed-area land use. There is also a general requirement that 10% of the energy for buildings must be from renewable sources. In addition, Yokohama promotes the acquisition of CASBEE certificates for the environmental performance of buildings. In the future, exploring the feasibility of using renewable energy will be a requirement, with regulatory steps introduced to ensure compliance.

To inspire environmental-friendly behavior in its citizens, Yokohama is also launching a new initiative called the Yokohama Environmental Point System. In 2009, as part of the 150th anniversary of the opening of Yokohama Port, the city will hold events promoting energy efficiency in individual households, educational programs aimed at elementary school students, the revitalization of local shopping districts, and grassroots cooperative efforts. After evaluating the results, beginning in 2010 the city will implement systems that encourage behaviors – like purchasing energy-efficient appliances – that reduce greenhouse gas emissions.

---

[5] Comprehensive Assessment System for Building Environmental Efficiency, an assessment tool developed by the Japan GreenBuild Council and the Japan Sustainable Building Consortium.

Since a major emphasis of the Yokohama efforts is on empowering citizens to reach environmental goals, the city has a host of programs implemented involving the use of renewable energy through "citizen power." The aim of these programs is the expansion of renewable energy usage to more than 10 times the current level (renewable energy as a part of total energy consumption in 2004 was 0.7%, and the goal is to push it to 12% by 2025). On the supply side, the program known as "Yokohama Green Power" involves the supply and expansion of the use of renewable energy through funding by citizens or by the sale of green energy certificates. The environmental value of the power will be bought at a fixed price. On the demand side, the city utilizes the information and technology from water treatment facilities in the area to make the bay area a high-tech center for renewable energy activities, with the aim of installing renewable energy or high-efficiency facilities in all city buildings.

In addition, Yokohama is instituting traffic measures to lower emissions from cars by promoting the introduction of electric and plug-in hybrid cars. For example, when trading in an old car for a new electric or a plug-in hybrid, one could receive a lower interest on the loan. The fixed property tax on infrastructure for recharging stations will also be lowered. The city will take the lead in promoting the technology through shared-use vehicles showcasing the "cool" design.

Finally, Yokohama also plans cooperative efforts with farming and mountain villages through the establishment of the "Big city – Farming and Mountain Village Cooperation Model" (*daitoshi nousanson renkei moderu*). Through this model, the city is cooperating with surrounding communities, like the small village of Doushimura, to establish systems that enable carbon offsets for organizations that support the administration of forests.

Other Ongoing Efforts in Yokohama

Even outside its activities for the Environmental Model Cities program, Yokohama continues to develop new policies to promote sustainability with an emphasis on individual involvement. Most recently, on December 12, 2008, the Yokohama city council passed a bill instituting a new tax to fund green initiatives called the "Yokohama Green Tax." Originally included as one of Mayor Hiroshi Nakada's policy proposals during the 2006 election, the municipal tax will collect 900 yen from individuals and 4500–270,000 yen[6] from corporations annually for 5 years, beginning with the 2009 fiscal year. Revenue from the tax will go toward the public acquisition of forest and farmland, the promotion of forestation, the maintenance of public land – including forests, and a policy promoting cooperation and volunteerism.

---

[6] Around $10 per individual and from $50 to 3000 per corporation (as of January 2009, when 1 yen is slightly more than $0.01).

Cooperation between industry, government, and academia has contributed to the success of Yokohama's environmental efforts, while the city's "citizen power" initiatives empower individuals to help remake their city a better place. The achievements of Yokohama prove that even a huge metropolis can be transformed into an eco-friendly community through a cooperative environment and the active involvement of its citizens.

The Zero-Emission House: Showcasing Japanese Green Technology

The future of sustainable residential living can already be seen first-hand at the Zero-Emission House, a model house built by the Ministry of Economy, Trade and Industry to advertise Japan's outstanding energy-efficient and environmental-friendly technologies. Though METI was the main sponsor of the project, co-sponsors included NEDO and the National Institute of Advanced Industrial Science and Technology (AIST), and companies all over Japan contributed technology to be featured in the house. Originally the house was built for display at the G8 Summit in Toyako, Hokkaido, Japan, where world leaders met to discuss vital international issues, including the prevention of global warming. The floor space is approximately 280 m^2, with a residential section taking up about 200 m^2 and a foot spa section the other 80 m^2. From the outside, it appears just like a traditional single-story Japanese home, but the technology on display encompasses almost every conceivable area of sustainable living.

The house itself is a lightweight steel-framed prefabricated house with a vibration control system and exterior walls covered with a dirt-resistant coating, reducing the resources necessary for both construction and upkeep, yet ensuring earthquake resistance. Environmental-conscious construction materials are used throughout, including materials reclaimed from thinned wood and waste wood. For power, the house features photovoltaic systems, a small wind turbine generator, several residential fuel cell systems, and a portable high-capacity lithium ion battery supply.

The interior of the house incorporates organic LED lighting (OLED) along with energy-efficient versions of appliances like televisions, refrigerators, washers, and dryers and has a home energy management system for keeping track of energy usage. A mirror duct system transfers daylight into the building for use by lighting fixtures, eliminating the need for electricity or thermal energy. A high-efficiency heat pump hot-water supply system provides hot water. A ventilation system filters out pollen and other allergens from the air while simultaneously using thermal exchange to bring external air to room temperature prior to intake. Special building materials in the walls regulate humidity and absorb pollutants and odors without requiring any energy, further reducing the load on the energy-efficient air conditioning units installed throughout the house.

By representing all the many ways houses could incorporate green technology, the Zero-Emission House makes it clear how sustainability and environmental action do not have to involve huge international or governmental efforts – they, quite literally, begin at home.

Conclusion

Japan's movement toward sustainability extends beyond local and even large-scale domestic efforts to the international arena. As the site of and signatory party to the Kyoto Protocol, Japan is committed by 2012 to reducing its CO_2 emissions by 6% from their 1990 baseline. Building on this existing commitment, in 2007 the then Prime Minister Shinzo Abe announced his "Cool Earth 50" initiative, arguing for a 50% reduction in greenhouse gas emissions globally by 2050 [16]. To achieve this lofty goal, he pledged that Japan would contribute cutting-edge technology in a variety of categories, including technology to reduce emissions from coal-fired power plants, enhance the safety and reliability of nuclear power generation, and reduce the cost while increasing the efficiency of solar power generation.

Recognizing the importance of involving all countries of our interconnected world in environmental efforts, the initiative also incorporates a "Cool Earth Partnership" to provide support to developing nations that want to make efforts to reduce greenhouse gas emissions. This way, while leading the way with cutting-edge technology and a variety of independent, integrated efforts at the domestic level, Japan can ensure other nations are not left behind. By utilizing technology, governmental incentives, corporate innovation, and grassroots cooperation, Japan will continue to make progress in sustainability and hopefully inspire other countries, cities, and communities to follow its verdant path.

References

1. METI Report on Eco-Towns: http://www.meti.go.jp/report/downloadfiles/g31118b50j.pdf
2. Kitakyushu Eco-Town website (Japanese): http://www.kitaq-ecotown.com/eco-town/1-2.html
   Kitakyushu Eco-Town website (English): http://www.kitaq-ecotown.com/about/english/
3. Kitakyushu *Eco-conbinato* website (Japanese): http://www.kitaq-ecotown.com/eco-conbi/a/index.html
4. Eco Premium section of Kitakyushu City website: http://www.city.kitakyushu.jp/pcp_portal/PortalServlet?DISPLAY_ID=DIRECT&NEXT_DISPLAY_ID=U000004&CONTENTS_ID=6650
5. Eco Action section of Kitakyushu City website: http://www.city.kitakyushu.jp/pcp_portal/PortalServlet?DISPLAY_ID=DIRECT&NEXT_DISPLAY_ID=U000004&CONTENTS_ID=14104
6. METI report on Next-Generation Energy Parks: http://www.enecho.meti.go.jp/080710energy-park.pdf
7. Japan Times article: "METI plans subsidies for home solar power" (06/25/08) http://search.japantimes.co.jp/cgi-bin/nb20080625a4.html
8. Reuters article: "Japan to bring back solar subsidy for homes" (12/24/08) http://www.reuters.com/article/environmentNews/idUSTRE4BN1U820081224
9. NEEDO summary of the Ota City Demonstrative Project: http://www.nedo.go.jp/english/activities/portal/gaiyou/p02050/p02050.html

10. NEDO presentation on the Ota City project: http://www.nedo.go.jp/english/publications/brochures/pdf/ota-project_nedo.pdf
11. Nikkei BP.net article on the megasolar Ota City solar project: http://eco.nikkeibp.co.jp/style/eco/special/081104_mega-solar04/index.html
12. Super Eco-House page on Ota City homepage: http://www.city.ota.gunma.jp/005gyosei/0090-001kankyo-seisaku/matimodel-h16.html
13. Report from Prime Minister's office on Environmental Model Cities program: http://www.kantei.go.jp/jp/singi/tiiki/080722kankyo-kouhyo.pdf
14. Yokohama G30 homepage: http://www.city.yokohama.jp/me/pcpb/g30/
15. CASBEE information: http://www.ibec.or.jp/CASBEE/english/index.htm
16. "Invitation to 'Cool Earth 50'", speech by former Prime Minister Shinzo Abe http://www.kantei.go.jp/foreign/abespeech/2007/05/24speech_e.html

Conclusion and Summary: The Next Steps

America has *no* energy policy today, let alone one that is sustainable and environmentally sound.

The USA and the world need to focus on that issue and what to do about it.

Only the EU (and not all countries comply with an energy policy), Japan, and the People's Republic of China have national energy policies, programs, and financing. However, none are progressive or proactive about sustainability and the environment. The need for a consistent, stable, and long-term energy policy stabilizes the energy markets along with creating jobs and business opportunities, since companies see and understand the multiyear policy as a predictable "road map" in that and related sectors (infrastructures like water, waste, transportation).

In other western nations, the shift from a fossil fuel and non-nuclear power-based economy (the Second Industrial Revolution) to a renewable energy-centered economy (the Third Industrial Revolution) has already begun. Japan and the EU are well into the process, since the millennium, to become energy independent through developing and producing renewable energy generation, storage devices, and other technologies. Their public policies and actual programmatic work is well documented [7].

The 2008 presidential election focused on the "economy." The choice was clear: keep the broken old neoclassical economy (Second Industrial Revolution) or begin the new one (Third Industrial Revolution). The new global economy has its philosophical roots in what some philosophers and scholars characterize as "social capitalism" [3, 4]. This next economy will be a "green" one that is focused on sustainable development for the public good in terms of its social concern for infrastructures like energy, transportation, telecom, waste, buildings, and natural resources. The 2008 presidential candidates and parties, all agreed that America must become "energy independent." But there was and is a vast difference in what those terms mean to each candidate and public party.

"Energy independence" must mean through renewable energy generation and be enacted as an American National Energy Policy to stop global warming and mitigate climate change. Energy independence does not mean drilling for more gas and oil in

---

Doug Grandy and Calvin Kwan assisted with the concluding chapter.

the USA or off its shores. And by using "clean technology" and "clean energy" such as natural gas and coal or nuclear power from within America's own borders. The "next global economy" is just that – a dramatic economic paradigm change from the neoclassical economics of energy (fossil fuels and nuclear power) to the renewable energy (green technologies) that are becoming more and more commercially viable and cost effective today as consumers install solar and wind energy systems at their homes, offices, and farms.

In the old neoclassical economic paradigm [3], there are long costly time periods to discover and drill for oil and natural gas that is then processed and shipped, transmitted, or piped to long distances to the customer. Coal is not a renewable energy resource, and it never will be "green." Coal is often touted (promoted by the industry today) as "clean," meaning that there are technologies that can make it cleaner then 100 years ago. However, there are still pollutants in the air such SO_x, NO_x, and particulates that are emitted and wastewater dumped into waterways that permanently damage the environment locally and migrate through the atmosphere to other parts of the region and world.

The same is true for nuclear power plants because the nuclear waste problem has not been solved in the USA, let alone France, which has 83% of its power generated from nuclear power. The capital cost for these facilities is beyond any accurate estimate and has, in fact, bankrupt one public utility in Washington state about 10 years ago. Currently, the state of South Carolina (Electric and Gas Company) started in the summer of 2008 to invoke a tax on all the utility customers in order to raise over $9 billion for the construction of one nuclear power plant producing only 1 MW of energy. Aside from the issue of waste from that plant, the actual stranded construction costs will take decades to recover, if at all. In short the same pattern and problems faced in Washington state are being repeated. However, a more basic issue from USDOE data is that there is only 61.5 years more of uranium available for current nuclear power plants. Imagine the storage and demand conflicts over uranium that could emerge should more nuclear plants be built now. Global conflicts will become even more hostile and severe.

The distinction, definition, and meaning of words like "energy independence" and especially "green" versus "clean technology" are critical for public policy makers and the general public to understand (Clark and Fast, ibid.). What has become even more significant is leaving the solution to energy production under the economic theory that there is no need for an energy plan and markets can set costs and prices for energy. However, little concern was raised about the impact on climate change that "de-regulation" (Europeans called it "privatization" or "liberalization") or "market forces" would cause, as California and other nation states discovered in early 2000. Even more significant, and as shown by way of examples throughout this book, technologies need to be integrated or even hybrid that leverage off one another in terms of invention, research, design, and demonstration in order to make them commercial and market ready [6].

Historically created public energy monopolies, such as utilities, controlled power generation from one location or centralized supply source, known as a "central power grid." When these utilities were "deregulated" in California and 19 other states, private companies took over and created "private monopolies" that bought the power-generated resources and controlled the energy supplies. The results were disastrous, since the same central-grid economic paradigm prevailed, but now as private monopolies. Government oversight, let alone regulations, was almost gone entirely.

Agile energy systems that are sustainable[1] provide a different energy and economic infrastructure from that of the current centralized grid systems. Furthermore, agile energy makes economic sense in the long run when compared to the associated costs of modernizing and replacing (let alone the issues of security and climate problems) transmission and distribution networks of the existing system. For example, in May 2008, the US Department of Energy (USDOE), Energy Information Agency, projected that electricity demand in the USA is to grow by 39% between 2005 and 2030. The USDOE also projected that at least 20% of the increased demand can be met using wind resources. The USDOE estimates that total capital cost to meet the 20% wind scenario would be nearly $197 billion. However, electric utilities would save at least $155 billion by not having to purchase as much coal and natural gas for conventional power plants. That would bring the net cost of the 20% wind power scenario to about $42 billion (EIA, ibid.).

This statistic does not take into account the added savings for not having to construct costly transmission lines. However, as in Northern Europe, when wind turbines are installed in towns and villages or remote areas, there is no need for massive transmission lines along with their high costs. Denmark, for example, plans to have 50% of all its energy from wind power by 2020. By 2009, it has already almost achieved that goal with close to 38% wind power generation from a combination of centralized wind power farms (including some in the ocean) along with local on-site wind power sites. This is an example of an "agile energy system" that has both wind farms and local wind turbines generating power.

Governments and countries all around the world are realizing the economic and societal benefits of constructing local, agile energy systems as opposed to just traditional transmission networks. Aside from the economic benefit of not having to build transmission lines, independent agile energy systems protect consumers against unexpected malfunction or shutdown of transmission lines. For example, in August 2003, untrimmed trees in Ohio became tangled up with central-grid power transmission lines, resulting in extended power outages in the northeastern USA and southern Canada impacting about 50 million people. Subsequent studies have concluded that a few 10 MW of distributed solar PV power could have prevented the blackouts. Agile energy systems can help prevent such system-wide blackouts as well as provide for emergency and security local power when needed.

---

[1] An "agile sustainable energy" system can be compared to the Internet (no central master computer mainframe but instead individual systems) in that they combine local on-site power generation systems of all kinds with those from the central power grid together. The concept also includes agile energy systems and infrastructures as well as on-site power, distributed energy systems, and combined heat and power.

For charts and graphs, see Appendix A.

The Problem Contains the Solution

The generation of power today from the fossil fuels and nuclear power costs is in billions of dollars. What most fossil fuel companies do not reveal are the actual costs for the power generated. That is, there are additional actual costs for the discovery of oil and gas locations along with the drilling and transmission, which is usually at least $1 million per mile but can add billions (especially along the ocean floor) to the actual power generation costs for electricity. Transmission usually results in a one-third loss of power sent over long distances. For consumers, the transmission costs add from 5 to 25% more charge per kilowatt-hour charged by the utility companies (PG&E, EDA, and Power Markets Week). Additionally, national governments give oil and gas companies high incentives and tax breaks. Whenever actual costs (including externalities) are measured, the consumer costs are considerably higher by a factor of four or more.

The "Next Global Economy" today [1] is the generation of energy for vehicle and stationary use through the renewable power generation of sustainable resources found locally such as sun, wind, geothermal, oceans, and rivers. There are no drilling costs and very little construction, operation, and maintenance costs. In fact, the costs for wind turbine power today are lower around the world than that for natural gas, after it has been drilled, piped, and processed for consumption. As noted above, far more significantly and rarely calculated in the costs for fossil fuels is the transportation, transmission, or pipeline flow, which can match even the initial discovery costs or more for these fuels. While there are transmission costs and accounting for the intermittency of renewable energy (especially solar and wind), these are far smaller than actual fossil fuels.

Herein lies the advantage of renewable energy for agile system (on-site power generation). Solar power as well as geothermal, wind, and biomass (recycled and waste materials) can be generated and used in local communities. There are no costs for securing the wind or the sun power. Furthermore, with local renewable energy generation, there is no need to be transmitted or piped from a central power plant to distant consumer locations and communities. Some utilities, like Southern California Edison, have taken the lead in seeking local, on-site power generation from solar systems placed on storage facilities and other large roofs whereby the power generated can be sold to local consumers.

Unlike traditional power generation, localized agile energy systems, such as solar PV, can be situated near load centers (Transmission Chart from USDOE, EIA). Generating at the point of use avoids the usual grid-based costs such as supply costs, transmission and distribution costs, and the cost of energy loss. Transmission and distribution costs can be two to three times the cost of bulk power (Shaw). Simply utilizing agile energy systems can result in estimated realized savings of $30– $50 per MWh simply from the avoided transmission costs (Beach and McGuire).

The result of local and central power generation is the creation of "agile energy systems" (*The Economist*, May 2004; [2]) that are sustainable and environmentally compatible with the local community. These "agile sustainable communities" thus use the combination of on-site power generation (such as the solar power on roofs or wind turbines that are specifically designed for buildings) and central-grid power systems for power generation. The local generation of power from solar and wind generation, for example, allows public and government buildings like city hall, fire and police, public schools, and universities to generate their own power, but still stay connected to the central power grid [6]. Increasingly, more and more private businesses have been following the public sector leadership.

The USDOE has numerous studies that document this fact. And several recently peer-reviewed papers do the same with the additional economic facts of savings in terms of short- and long-term energy and fuel costs. Consider these facts from the US DOE in terms of Quads of BTU consumption in 2005 whereby there will be 99.895 consumed in fuels and power use, whereas by 2030, over 131.168 BTUs will be consumed at the current level of consumption and without significant renewable power generation supply. When the local or state level is considered, California as an example has 6.7 Quads demand in 2005, but with its projected almost double population growth, it will have 6.4 Quads by 2050. As one independent academic study (Matheson, 2008, unpublished) expert noted, California can replace all its heating, transportation, and electrical energy needs primarily from fossil and nuclear power fuels (about 75% not including hydroelectric power) with renewable energy generation over the next 45 years.

The problem of global warming starts at the local level because people, business, and groups demand secure energy for their personal and professional needs. However, historically, the public is a victim of limited choices for energy in their homes and fuels for their cars. For the consumer, energy came from a central grid that was usually powered by fossil fuels. Published research evidence has shown how the USA auto industry sought to force consumers into buying internal combustion engine cars (e.g., that use fossil fuels) by marketing to the public and forcing governments to shut down or restrict mass electric rail transportation. The classic case was the conflicts in San Francisco over retaining its Trolley Car System or removing it in favor of wider roads and freeways. The city won that battle in the 1950s and a later one to establish the Bay Area Rapid Train System (BART) in the late 1960s against extreme pressure from the American automakers.

Then 300 miles south, in the 1920s, the Los Angeles area once boasted one of the most extensive electric rail systems in the country. By 1945, again by pressure from automotive manufacturers, the rail systems were replaced by buses and internal combustion engine cars. Since then, traffic and congestion in Los Angeles have grown immensely. Efforts to revive and rebuild a Los Angeles electric rail system have been derailed by the immense costs.

Above all, getting to any kind of solution requires the combination of government (civic) and business (markets) collaborations working together [5]. These "civic–markets" must define sustainable development goals that protect the environment while meeting consumer demands for energy. Rules, standards, and codes must be established with the environment being the top priority. There is no neoclassical economic balance when it comes to protecting the environment versus economic development. The latter concern with business, jobs, and growth can be directed by the former, where the future of the planet comes first and foremost.

And therein is the solution. The Chinese have one word (wiejei) for both the "problem" and the "solution" in almost any situation. Local renewable energy generation for homes and businesses can come from the geothermal, sun, wind, and ocean power among others, whose "fuel" costs do not exist. The wind and sun are there as part of natures resources. Hence, they are sustainable, and when in combination with energy storage devices such as fuel cells, batteries, and hybrid technologies such as pump storage, etc., they meet basic consumer demands that mitigate global warming and climate change. Creating local energy generation is environmentally sound and creates economic development and jobs, but above all provides energy supplies that can reduce local air emissions and stop pollution that impact the world communities.

Agile Sustainable Developed Energy Systems

"Agile energy systems" are flexible and adapt to change effectively and efficiently for economic, environmental, and social benefits or what is known as the triple bottom line. Recent statics reveal that with the downturn of the USA economy in 2008, the private sector building and construction industries were impacted. However, by the third quarter of 2008, there has been a reversal with new government building and construction now leading by over 10% the private sector in terms of buildings, infrastructures and clusters of facilities. Now communities, cities, and nation states are making plans and public policies to create the "government market" for procurement and coordination of public resources for renewable energy on-site and central-grid power generation. Agile energy systems, ideas, and areas of regional organization and structures are diversified and use renewable energy sources which thus make agile systems less vulnerable to disruption and more reliable.

Another area is the balance for agile systems, since they emphasize the best use of energy, not just the amount of power available. Balance involves promoting conservation, encouraging shifting of use to non-peak times, and reducing the consumption, not just selling power. Most importantly, there must be local or distributed generation along with redundancy or backup from the central grid. Agile systems are therefore independent yet interconnected to the power systems in regions. Hence, they find ways to avoid bottlenecks in delivery power that integrates energy, which are traditionally separated. For example, cogeneration integrates electricity and heat systems. Hydrogen from renewable energy sources can pair technologies: for example, wind- or solar-produced electricity can be stored in hydrogen fuel cells for transportation and then, when not used for fuel, can be used to generate stationary power for buildings.

Agile energy systems are smaller, on-site, or distributed in locations close to where energy is needed and coupled especially with renewable resources. Neighborhood-scale systems are ideal because they recognize environmental costs and can be linked to the grid. Hence, an agile system links the community to other organizational levels such as regional, national, and even global areas while being aware of the public good as a primary goal. Government, at all levels, needs to implement public policies that create and oversee "agile communities." In short, governments must recognize the diversity of communities in terms of not only energy needs but also their sources of energy production.

In every region of the world, there are different strengths in sources of energy. Global and world priorities must be, however, on renewable energy sources for local energy generation. Climate change and global warming now clearly dictate that priority. Both government and industry need clear, concise, long-term-oriented, and consistent market rules, standards, codes, and operating protocols in order to achieve the goals of sustainable society and business. Hence, the "civic–markets" adhere to five principles, which are outlined for any such energy framework:

1. *Government oversight of existing utilities.* The traditional regulatory framework is close to micro-management as all aspects of the utility are regulated – for example, prices are set and specific technologies are mandated. A new model of oversight would not replace existing utilities or micromanage them but help them move toward being agile systems by the right combination of incentive and mandate. One example is the creation of Renewable Portfolio Standards (RPS) that have targeted contracts by the state for renewable power and distributed power. Another is working on helping intermittent power producers (especially wind and solar) to use the grid for backup, security, and storage for obtaining intermittent power so that they act as "batteries" and hence made intermittent power into base load suppliers.

2. *Strategies for transitional costs.* The transition from the regulated systems to the agile sustainable energy system needs the tools for dealing with unknowns and negative unintended consequences. The energy crisis in California cost the state at least $40 billion in part, due to the fact that the state signed long-term energy contracts at inflated prices, which cannot legally be rewritten. Innovation is more difficult due to these huge stranded costs. The state in 2008 faced a large budget deficit, with much of funding relying upon "bonds." A number of bond measures advocated by wealthy proponents of fossil fuels (including oil and natural gas companies) along with pressures for nuclear power plants will make the so-called "transitional costs" into decade-long liabilities. Natural gas, even if produced in the USA, will require transmission via transportation, pipelines, and extensive power generation plants. It will take decades for the costs to be recovered through bonds. So "transition" means stranded costs from 30 to 40 years or more. And the impact will be on the general public and middle-income consumers. The scenario for nuclear power plants is even worse, and to call them "renewable" is a deliberate misuse of the word and concepts behind renewable energy power generation.

3. *Joint public–private investment.* Known as "civic–markets" [7] the combination of public and private partnerships includes investments and finance. Public ownership may not be the right solution, but the public sector must stay involved.

Investments from public sector pension funds, for example, would allow public influence in the utility through voting membership on the Shareholder Board and executive committees. Another example concerns the government and public sector setting leadership standards in "green public buildings" requirements. California has started this process with the setting of LEED (i.e., leadership in energy and environmental design from the US Green Building Council in Washington, DC) standards for public state buildings. These approaches to buildings and development are sustainable and have a positive impact on reducing pollution in the environment.

4. *Innovative renewable technological systems.* Governments can develop market mechanisms to direct the market to innovative technologies through new regulations, codes, and standards. For example, sustainable building codes can factor in energy and material efficiencies. Procurement is another mechanism that allows the government to lead what commercial and business sectors should be doing as well. The government can set requirements for buying hybrid cars or all-electric vehicles. These requirements can be posted to industry for "master contract" purchasing of goods and services. However, the most significant government area for leadership is in the accounting and economics of sustainable energy supply. Through the use of "life cycle analysis" accounting, government can require that long-term variables need to be considered for how energy is produced, transmitted, and distributed. Taking into account factors like health, the environment, and even basic processing costs for drilling oil and gas will result in the actual costs for fossil fuels. Again, learning from Europe, Germany enacted "Feed-in-Tariffs" over two decades ago that supported the growth of wind power generation and then increased the tariffs so that solar power would be dramatically increased. Today, Germany is the world leader in solar power installations and manufacturing.

5. *Agile systematic and regional implications to the power system.* Opportunities for renewable energy include sustainable economic development such as jobs and business development from emerging technologies that produce increased employment in "green" energy and environmental businesses. Environmental considerations must be part of any energy policy and plan. Once these policies are created, they must be implemented with the environmental impact in mind at all phases. Buildings require that the use of materials such as furniture and supplies should come from recycled resources. The connections and infrastructures between buildings and within them must be from renewable energy sources. Each energy use area must be agile in order to allow for flexibility and change.

References

1. Clark W W. II Lead Author, "California's Next Economy", Governor's Office Planning and Research. Sacramento, CA 2003.
2. Clark WW. II, Bradshaw T, *Agile Energy Systems: global lessons from the California Energy Crisis.* London, UK (Elsevier Press, October 2004).

3. Clark WW. II, Fast M, *Qualitative Economics: Toward a Science of Economics*. London, UK (Coxmoor Press 2008).
4. Clark, Woodrow W. II and Xing Li, "Social Capitalism: transfer of technology for developing nations", International Journal of Technology Transfer, Inderscience, London, UK, Dec 2003.
5. Clark, WW. II, Lund H, "Civic markets in the California Energy crisis", International Journal of Global Energy Issues. Inderscience 16(4):328–344, Dec, 2001.
6. Clark WW. II, Lund H, "Sustainable development in practice", Journal of Cleaner Production, Elesvier Press, 15, 253–258, Dec 2006.
7. Rifkin J, The European Dream. The European Dream: how Europe's future is quietly eclipsing the American Dream, Penguin, 2004.

Appendix A
Los Angeles Community College District (LACCD)

Tony Fairclough, Andrew Hoffman, and Larry Eisenberg

The 21st Century "Green Energy Economy": The Third Industrial Revolution

Agile sustainable communities have flexible infrastructure systems that are able to change quickly. An agile sustainable system can adapt due to change, where innovation is welcomed, rather than opposed. These systems are resourceful and are adept at developing ways to avoid or solve conflicts, while deconstructing social–economic barriers that slow down effective solutions to problems. See the chart of an "Example of an Agile Smart Energy System" in the Introduction. These systems are integrated or hybrid and thus "smart" because of the use of wireless and Internet, which also have higher energy costs but now offset with the onsite power generation from renewable energy sources, for the individual homes and customers of all kinds (Clark and Eisenberg, 2008).

Agile systems are dynamic and progressive since they are not limited and inhibited by either political pressures or constraints from lobbyists and special interest groups. Instead, such agile systems foster and promote diversity at the local level, along with dynamic growth from which there is change using knowledge, intellectual capital, financing mechanisms, and advanced technologies such as wind power, geothermal, solar thermal, photovoltaic, CHP (combined heat and power), fuel cells, and hydrogen options together with conservation and load management.

All these and other integrated or hybrid technologies were discussed throughout the book in the light of being components in agile energy systems. Here, however, are some examples and cases of technologies that are now commercial for sustainable communities. More are being developed all the time. This sector has become prominent and seen as one of the key economic growth engines for all nations as the international communities recognize the need for stopping global warming and climate change through renewable energy power generation for buildings and transportation.

However, there is not one standard, set, or uniform set of circumstances or technologies that fit all communities and regions. There are needs for uniform integrated technologies as well as codes or rules for performance, operations, and maintenance. Geothermal is not found everywhere. Nor are sunshine and wind, but the combination of these renewable resources along with new technologies for storage such as fuel cells and flywheels provide for hybrid technologies that create firm base loads and dependable power. Such systems can replace or complement conventional fossil fuel system or be ready in the transition to fully energy-independent on-site and distributed energy systems.

Agile sustainable energy systems are a combination of local and regional infrastructural systems (including on-site, distributed, and self-generation) which include the central water and energy grid systems and can be used in the future, primarily for redundancy, storage, and backup purposes. These agile systems are not particular technologies or market mechanisms, but rather the Third Industrial Revolution paradigm shift that can be described as a new civic-market orientation. Such agile sustainable systems for communities have the following features that help establish and guide technological innovation, demonstration, and commercialization:

- *Integrated diversity:* Agile sustainable systems use diversified and hybrid integrated renewable energy sources that make them less vulnerable to disruption and more reliable, especially because they are less reliant on distant suppliers who are dependent on declining fossil fuels. This diversity also builds in needed redundancy.
- *Economic balance:* Agile systems emphasize best use of energy, not just the amount of supply and demand. The balance involves promoting conservation, encouraging shifting of energy use to nonpeak times, and reducing consumption.
- *Interdependence and interconnection:* Agile systems find ways to avoid bottlenecks in delivery and integrate energy usage, which are traditionally separated. For example, cogeneration integrates electricity and heat systems. Hydrogen from renewable energy sources pairs technologies: for example, wind-produced electricity can be stored in hydrogen fuel cells for stationary and transportation fuel.
- *Spatially appropriate:* Agile systems are smaller and located on-site for buildings but also for vehicle power. They should locate close to where energy is needed and coupled with renewable resources. Building specific or on-site generation is critical. Neighborhood-scale systems or distributed generation is ideal because it factors in environmental costs, but can be linked to the grid to act as storage for intermittent times of use as well as supply the central grid with renewable energy generation.
- *Community, regional, and nation state:* An agile system links the community to regional, national, and global societal levels. The agile system is the infrastructure that transmits energy but also supports other infrastructures like transportation, telecom, water, and waste.
- *Social and public good:* Agile systems have people and the public good as their primary goal. This is called the "civic core" or the need for the energy market to be government driven for the local community and citizens.

In short, agile sustainable systems make up the basic infrastructures for the new energy economic paradigm of the 21st century that makes every community energy independent and carbon (particulates and pollution) free. Stopping climate change and global warming must start at the local community level. Agile sustainable systems are needed in every home and mode of transportation for work and leisure activities. Soon (within the next 5–8 years) they will be the norm. Our planet has no other choice but to change dramatically.

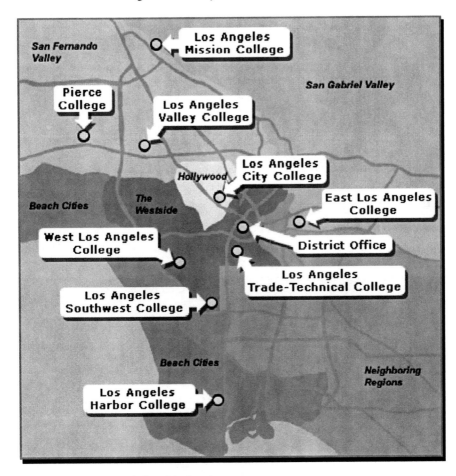

Los Angeles Community College District: nine campuses

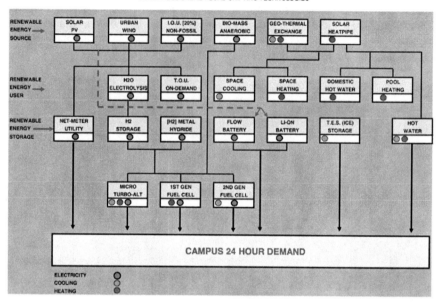

LOS ANGELEC COMMUNITY COLLEGE DISTRICT
RENEWABLE ENERGY LOAD SHIFTING TECHNOLOGIES

Base load data

Audubon Center in Debs Park in East Los Angeles. Tel-Learning Center totally off the grid through the use of renewable energy technologies

Solar heat pipe technology used for heating/cooling at Audubon Center

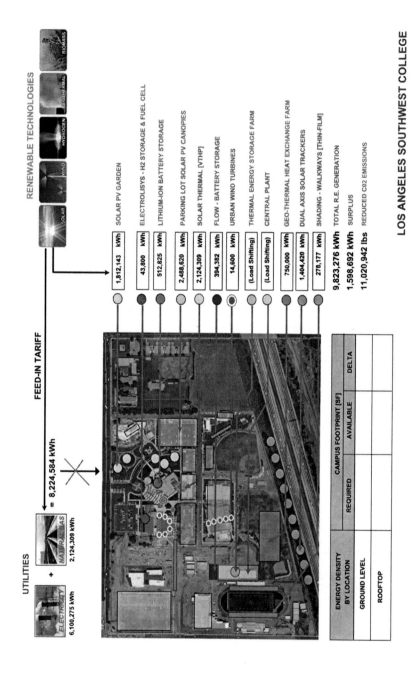

Renewable energy and masterplan for LA Southwest College

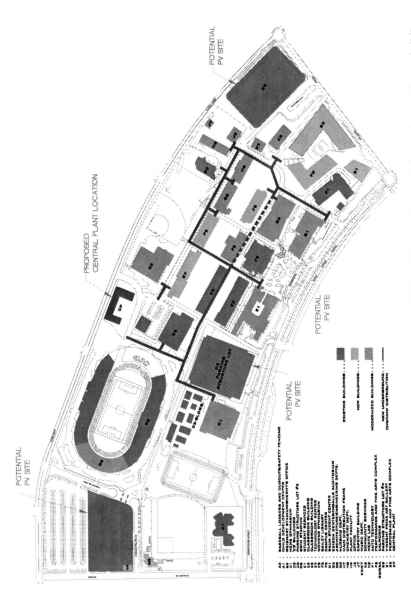

East Los Angeles College. Key: *Red areas* are solar/PV installations (about 4 MW) and *blue area* is central plant and connections to buildings

Appendix B
Aragon and Navarra, Spain

Nick Easley

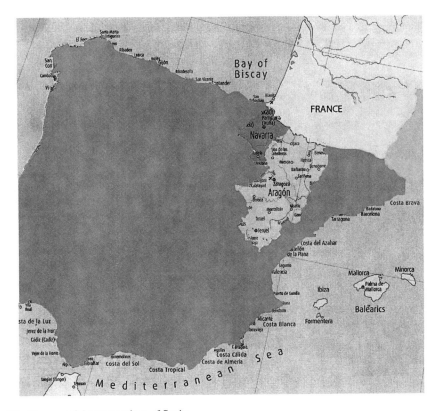

The Navara and Aragon regions of Spain

Aragon

Aragon, one of Spain's 17 autonomous communities, is located in northeastern Spain and is comprised of three different provinces covering around 47,000 km^2 of land. The population of Aragon is estimated at 1.2 million, with half of the region's inhabitants living in Zaragoza, the capital city. Regional GDP reaches 32.6 million euro.

Aragon is among the richest autonomous regions in Spain, with GDP per capita above the nation's average. The traditional agriculture-based economy from the mid-20th century has been greatly transformed in the past several decades, but service and industrial sectors are now the backbone of the economy.

The chief industrial center is the capital city Zaragoza, where the largest factories are located.

In 2007, Navarre generated around 60% of its electricity requirements by means of wind power and small hydropower stations.

GM Zaragoza

The world's largest rooftop array is installed on a General Motors assembly plant in Zaragoza, Spain. The project will be 10–12 MW, a huge number in a field where most arrays are measured in kilowatts, units 1,000 times smaller.

Solar array being installed on a General Motors assembly plant in Zaragoza, Spain
Source: http://www.autobloggreen.com/tag/gm+zaragoza+solar/

Walqa Technology Park

The Walqa Technology Park in Huesca, Spain, is nestled in a valley in the Pyrenees and is among a new genre of technology parks that produce their own renewable energy on-site to power their operations. There are currently a dozen office buildings in operations at the Walqa Park and 40 more slated for construction. The facility is run entirely by renewable forms of energy, including wind, hydro, and solar power. The park houses leading high-tech companies, including Microsoft and other IT companies, renewable, and so on.

The Walqa Technology park in Huesca, Spain
Source: http://www.mo-di.net/september_2006_dir/17/view

Navarra

Navarra is a smaller, autonomous community of Spain also situated in the northeast of the Iberian Peninsula and bordering Aragon to the northwest. The region is around 10,000 km^2, with a total population of around 600,000 (of which approximately one-third live in the capital, Pamplona). The GDP in Navarra is also above the national average at 17.7 million euro.

Acciona Zero Emissions Building

The headquarters of ACCIONA Solar in Navarra consumes 52% less energy than a conventional building with the same characteristics, as a result of the application of a large number of energy-efficient and energy-saving measures, meeting 48% of energy requirements with photovoltaic solar energy (electricity), thermal energy (heating), and biodiesel. It is the first office building certified as "zero-emissions," which means that all its energy requirements are met without any greenhouse emissions.

The headquarters of Acciona Solar, in Navaro, Spain, a certified "zero-emission building"
Source: http://www.pvdatabase.org/projects_view_details.php?ID=336

Other solar production sites in Spain:

http://www.industcards.com/solar-spain.htm.

Sources:

http://www.gm.com/corporate/responsibility/environment/news/2008/solar_070808.jsp.
http://www.smart-rfo.org/technologypark/download/WALQ%20 PRESENTATION%20(Smart).pdf.
http://www.agencyscience.com/article.php?q=08111401-novel-system-proposed-optimise-combined-energy-use.
http://www.acciona.es/corporate-responsibility/climate-change-and-eco-efficiency/ecoefficient-construction.aspx.
http://www.spainbusiness.jp/icex/cma/contentTypes/common/records/viewDocument/0,,,00.bin?doc=4159801.

Appendix C
City Design and Planning: A Global Overview and Perspective

Stephanie Whittaker and Steve Done

The population of the world is predicted to grow to 9 billion by 2050, with the urban population expected to more than double from 2007 figures to 6.4 billion by 2050.[1] As the urban population grows, there will be a rapid growth in resource consumption and CO_2 emissions; however, the current model for urbanization is unsustainable, and we need to identify ways to develop a more ecologically sound model which begins to address the problems we face. So, that lower-to-middle-income countries can urbanize in a way that makes the most efficient use of natural resources, and high-income nations can begin to tackle many of the issues that have led to an over-consumption of these resources. Arup's approach to eco-cities has been developed to tackle these issues and create a model that can be applied to any type of city in any location.

Arup believes that for urban centers to move toward an ecological age, there needs to be a shift from an economic model based around consumption to one in which a desire for resource efficiency is the driving force. A rapid paradigm shift is required to greatly reduce the demand for resources and, in parallel, identify more diverse and efficient renewable sources of supplies. To achieve this we need to reconnect urban and rural communities to close resource loops and achieve a sustainable future. This will mean radical changes in agriculture, manufacturing, urban systems, and in the way we live and the choices we make.

By their very nature urban centers are very complex, combining commercial activity with the residential and leisure needs of the population, and the infrastructure required to support this. Compounding this is the fact that many of the issues urban centers face are interrelated, and this means that creating solutions to reduce emissions and improve resource efficiency requires taking a wider view to deliver the best result. Arup's approach acknowledges this complexity and examines the relationships and interdependencies of stakeholders, institutions, uses, activities, and impacts, to establish solutions which have clear and measurable outcomes and benefits and which resolve the unique issues every city faces.

---

[1] The United Nations World Urbanization Prospects: The 2007 Revision, United Nations Publication.

We also recognize that creating a sustainable city is not just about using resources more efficiently and that for a city to be truly sustainable, a variety of factors need to be considered. Such factors include the need to create urban environments that people want to live in because they provide jobs and a good quality of life and the need to encourage behavioral change to achieve lasting results, either through policy and strategy development or through the creation of environments that facilitate the process of change, for example better public transportation, landscaping, city layouts. Not only this, Arup helps cities to understand if and how they could apply new technologies or existing best practices to achieve the most optimal results for themselves. This is key as not every technology delivers the same results in every area of the world, and best practices may need to be adapted to ensure success in another city or would fail to deliver anything at all.

The integrated thinking required to do this has always been at the center of Arup's approach. From the outset, our founder Ove Arup believed that groundbreaking solutions and unexpected synergies are often revealed when diverse disciplines work together as one. Within the urban environment this type of holistic integrated approach is key to identifying the real opportunities that will enable existing or new cities to take the steps toward the ecological age. Beginning with our work at Dongtan, Arup has employed this holistic approach to planning and designing sustainable cities. This is achieved by bringing together multidisciplinary teams whose work is shaped by an overarching development framework, which is determined at the outset of each project and typically covers transportation, utilities, economics, and implementation, and risk, social, political, and sustainable issues. The framework also establishes goals, objectives, and performance indicators so that rigorous attention is given to measuring the outcomes for each project. These then form an integral part of the design process and are incorporated into all phases of the decision-making process. By applying this approach we ensure that we achieve an optimal balance of priorities and create demonstrable outcomes, which allow cities to make the choices and implement solutions that will enable them to not only become sustainable but also realize the benefits that the move toward sustainability brings.

Developing the Tools to Create Sustainable Eco-cities

Arup has developed two tools which were created to specifically allow a comprehensive assessment of the impact of development scenarios and comparison of non-quantifiable parameters at all stages of the decision-making process.

Sustainable Projects Appraisal Routine (SPeAR®)

SPeAR® allows a qualitative assessment and comparison of non-quantifiable parameters. It is based on a four-quadrant model that structures the issues of sustainability into a robust framework, from which an appraisal of performance can be undertaken. SPeAR® brings sustainability into the decision-making process with its focus on the key elements of environmental protection, social equity, economic viability, and efficient use of natural resources. As such the information generated by the appraisal prompts innovative thinking and informs decision-making at all stages of design and development. This allows continual improvement in sustainability performance and assists in delivering sustainable objectives. SPeAR® is a unique framework that allows for logical structuring of the diverse issues of sustainability. It also provides a methodology for appraising the performance of the indicators and transferring the outcome into a diagram that gives a unique "visual" profile of sustainability.

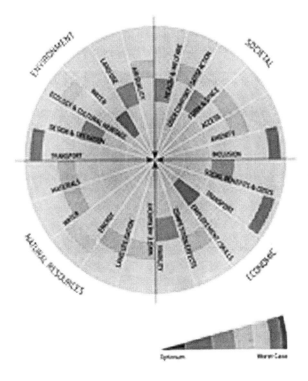

In addition SPeAR®:

• Highlights strengths and weaknesses and points to future areas of opportunity
• Provides management information to aid decision-making
• Provides auditable information for assurance and verification purposes.

Integrated Resource Management IRM

The Integrated Resource Management (IRM) model provides a comprehensive assessment of the impacts of development scenarios, including carbon emissions, social equity, and infrastructure cost implications. The model employs defined performance indicators (KPIs) and targets for a quantitative evaluation of plan proposals. IRM is a powerful tool in testing performance and informing design modifications throughout the process. The benefit of the IRM service stems from its ability to test selected masterplan options and scenarios in order to derive an optimized "preferred" solution through an iterative process of "assess-select-review-improve."

Results Generated by this Approach: Sustainable Eco-city vs. Conventional City

The planned results for Dongtan highlighted below show the benefits that our approach generates for cities when compared to a conventional city approach.

	Sustainable eco-city	Conventional approach city
Energy	Energy demand 600GWh/year No CO_2 emissions from energy for power and heat	Energy demand 1650GWh/year 350,000 tons of CO_2 emissions
Water	Water consumption 16,500 tons/ per day Water discharge 3,500 tons/per day	Water consumption 29,000 tons/ per day Water discharge 29,000 tons/per day
Transport	Daily travel 4.2 million km Zero CO_2 emissions Average trip length 24 km	Daily travel 6.0 million km 400,000 tons of CO_2 emissions/year Average trip length 56km
Ecological footprint	2.6 global hectares per person	5.5 global hectares per person
Creating jobs and homes	80,000 residents 51,000 jobs	50,000 residents 19,000 jobs

Other Examples of the Arup Approach at Work

Although Dongtan is perhaps the most well known of our projects, we have or are currently applying a similar approach to a number of cities around the world.

Stratford City, London, United Kingdom

The redevelopment of the Stratford Rail Lands in East London offers a unique opportunity to create a new metropolitan center based around the Channel Tunnel Rail Link (CTRL) International and existing Stratford Stations.

Working with Stratford City, and a team of architects and landscape designers, a masterplan has been developed for a new metropolitan centre for London. The new city will be integrated into existing Stratford Town and will incorporate 13.4 million square feet of mixed uses including retail, commercial, leisure, municipal, and hotel and up to 14 million square feet of residential.

A network of linked landscaped places will be incorporated to create the public realm for a high-quality urban environment. Spectacular buildings will be constructed which will establish the center as the major landmark site in the Thames Gateway. Improved and extended public transport networks, new roads, and pedestrian and cycle routes will link neighboring communities into the core of the development. The existing rail infrastructure is integrated to maximum opportunity.

The plan is complemented by a sustainability strategy dealing with issues of water, energy, waste, microclimate, air quality, and ecology, ensuring the delivery of short-, medium-, and long-term goals.

Dongtan Eco-city, Shanghai, China

Dongtan eco-city, set over 8,600 ha on Chongming Island near Shanghai, is a landmark project designed and master planned by Arup. With an unprecedented number of sustainable strategies integrated into every aspect of its design and implementation, Dongtan will be a model for low-carbon urban development in the future. Dongtan is expected to reach a population of 5,000 by 2010, 80,000 by 2020, and half a million by 2050. It is designed in the form of three compact villages, with the first phase, the 250-acre (100-ha) South Village, to include 2,500–3,000 dwellings, a portion of the marina, and open space and park

The need to protect local wetlands and the adjacent Ramsar bird habitat was a key driving factor for the design. The plan will restore and enhance wetland areas

Dongtan Eco-City, Shanghai, China © Arup

to create a significant "buffer-zone" between the city and the mudflats, 3.5 km wide even at its narrowest point.

Dongtan's design is based on a network of water features, canals, and lakes. Sustainability features include the following: alternative transportation including water taxi and hydrogen fuel buses, battery and fuel-cell powered cars, 7-min walking distances to public transport, employment hubs to minimize travel, electricity and heat generation entirely from renewable sources, a combined heat and power plant fueled by biomass, and 90% waste recovery and recycling.

Zuidas Development 2007, Amsterdam, the Netherlands

Zuidas is a district within the city of Amsterdam, which is characterized by a central corridor of overground infrastructure including rail and highway that acts as a barrier to seamless development in the area. The vision for Zuidas builds on current plans to submerge the existing infrastructure and creates dynamic new uses and amenities for the reclaimed space above ground. To create an urban identity that will be unique in the Netherlands, Zuidas will incorporate high standards for sustainable development. It will be a high-quality cityscape, a place where people want to be, and a model for flexibility and adaptation over time.

The central core accommodates 930,000 m^2 of residential, office, retail, and amenities around a new international station for high-speed trains and a robust network of underground infrastructure and access routes. The creation of a collection of

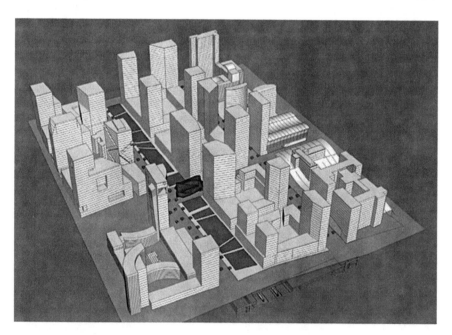

Zuidas, Amsterdam © Arup

new public squares as part of the open space network of Zuidas and its surrounding areas is the cornerstone for a timeless and long-lasting development framework for Zuidas. Looking beyond its immediate context and boundaries, Zuidas's street grid and open space system enhance connections with the historic city of Amsterdam and between the parks located to the east and west.

Jeddah Central District, Saudi Arabia

Jeddah, a city with 3.4 million people, is strategically located on the Red Sea on the west coast of the Arabian Peninsula. It is the commercial capital of Saudi Arabia and the gateway to the holy cities of Mecca and Medina. In recent times, the city's expansion has turned its back on the historic port and core and pushed outward along the coast. Burdened by a highway that bisects the site, traffic congestion, and contaminated water in the surrounding bays, the historic district is dominated by vacant and underutilized land.

The aims of the project are to regenerate the historic core and strengthen the city's economy and role as the gateway to the Arab world based on sustainable planning principles. An innovative clean water strategy for the bay was a first step and cornerstone for improving the waterfront and unlocking the historic city's development potential.

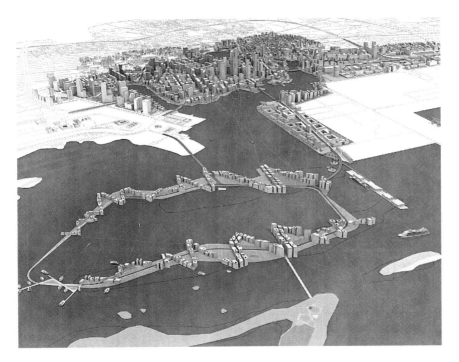

Jeddah, Saudi Arabia © Arup

Another important element of the plan was the development of key city corridors, a pattern seen in the historic core and across cities in the Middle East. The corridors are derived from the existing grid and strengthen connections between the city in the east and new developments in the west. The corridors will accommodate vehicle and public transport access as well as shaded public spaces and pedestrian amenities.

An iconic development proposal for the offshore barrier reefs of Jeddah is another feature expected to provide impetus for new waterfront development.

Waterfront Dubai

The new city of Waterfront in Dubai covers an area of 120 km^2 close to the Abu Dhabi border and is designed for a population of 1.56 million. It is targeted for completion in 2022. The key objective of the Energy Strategy and associated Urban Design, Water, and Building Guidelines is to substantially reduce energy demand compared to a business-as-usual (BaU) scenario for the development. The strategy includes the integration of renewable energy sources, resource management, and sustainable movement into the masterplan and increased building performance standards to achieve around a 60% reduction in energy demand compared to BaU.

Following an analysis of the existing masterplan, refinements were suggested to incorporate the Energy Strategy recommendations and a set of urban design guidelines were prepared to help achieve the energy demand targets. These include:

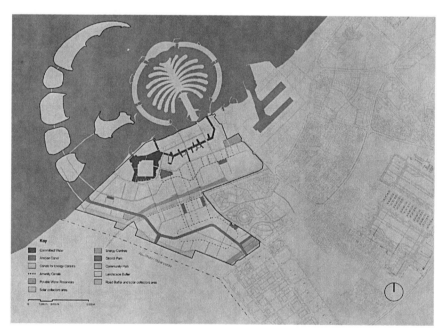

Waterfront, Dubai © Arup

- design features with a key functional role in the Energy Strategy such as large areas for solar collectors and logistics canals
- balanced communities with jobs and homes that minimize the need for travel
- improved access to transport, which includes routes through internal modified environments
- streetscape design such as planting shading and local waterbodies to improve conditions for walking and cycling

Urbanya, Santiago, Chile

The main objective was to review and analyze the 688 ha masterplan that has been developed to date, in order to incorporate elements of sustainable urban design. Arup's role in collaboration with local team is to integrate its sustainable approach in the design process of Urbanya to:

- improve urban design quality and integration of systems
- add value through a brand positioning
- add value through the reduction in operational and lifecycle costs both for operators and users
- improve the performance of the project in terms of overall energy and water utilization, waste resources, CO_2 emissions, and land consumption

These will generate additional value to the land and real estate products.

Urbanya, Santiago, Chile © Arup

La Spezia, Italy

La Spezia is a city in the process of change, looking to reinvent itself to become a major Mediterranean destination. Its central location in Northern Italy and Europe, year round good weather, historical town center, and good local services give it exceptional potential to grow beyond its original function as a port and attract new business, jobs, people, and tourism. The proposal for La Spezia aims to grow the city's scale of influence and allow it to develop into a vibrant and successful waterfront city.

The strategies for La Spezia to achieve its objectives focus on ways of improving existing infrastructure, targeting a wider group of users and harnessing the potential for leisure and tourism uses. Key principles include

- growing into a knowledge-based economy with an emphasis on innovative sustainable technologies
- recognizing the harbor as an asset not a barrier and promoting connectivity with Nervi, Porto Venere, and Cinque Terre with a solar-powered public transport network
- identifying underutilized harbor sites that can be redeveloped into mixed use, compact, and high-quality places
- generating a wide range of tourism and leisure activities for residents as well as visitors
- demonstrating clean technologies for regeneration, which in itself could show-case the city's new identity

La Spezia, Italy © Arup

Arroyo Seco, Los Angeles, California

Working in partnership with the Los Angeles Department of City Planning (DCP), a Specific Plan and Environmental Impact Report will be created for the 330 acre Cornfields/Arroyo Seco site between Chinatown and Lincoln Hill/Cypress Park

Metro Rail Stations in downtown LA. The site, viewed by city leaders as an opportunity to achieve a cluster of sustainable industry, infrastructure, and living will also support other related efforts such as the city-wide green building program, Los Angeles River Revitalization Plan, and the Mayor's Climate Action Plan.

Using the same integrated urbanism methodology applied to the Dongtan eco-city project, the plans for Arroyo Seco will provide a framework for the district's future redevelopment, infrastructure investment, and economic regeneration – creating a new standard for urban design and community living in Los Angeles.

Old industrial land will be redeveloped and retrofitted to attract clean technology businesses and create new green collar jobs. Public spaces will be created reversing the trend of a lack of parks and plazas in the city's neighborhoods, along with "complete streets" that accommodate pedestrians, bicycles, automobiles, and transit as well as trees for carbon sequestration. Moreover, the presence of the metro line running through the site and good links to public transport corridors will make the site easily accessible and encourage transit-oriented development.

Vista Canyon, Valencia, California

A sustainability framework was created which would provide the basis for integrating sustainability strategies into the Vista Canyon Ranch development in Valencia, California. The goal of the strategies was to allow the developers to improve the environmental stewardship and social benefits of the development as well as its economic viability.

The sustainability framework provides the Vista Canyon Ranch developers with the first steps toward achieving a practical, successful, and economically viable sustainable development. If the framework is implemented, it will help the developer create a unique sustainable vision for the community and allow them to embark upon the path to a development which is truly sustainable.

The work we carried out to develop the sustainability framework has now been included within the Vista Canyon specific plan which has been submitted to the City of Santa Clarita. The specific plan serves as the masterplan for the project area and serves as a guide to implement the goals, policies, and objectives of the city's General Plan. The intent is to establish a relationship between the land owner Vista Canyon Ranch LLA and the city to enable the creation of new development opportunities and assist the city in achieving a sustainable community with a very high quality of livability.

Index

Note: The letters 'f' and 't' following the locators refer to figures and tables respectively.

LaVergne, TN USA
05 December 2009
166077LV00003B/14/P